Math Mammoth Fractions 2

By Maria Miller

Contents

Introduction

Math Mammoth Fractions 2 continues the study of fraction topics after *Math Mammoth Fractions 1*. I sincerely recommend that the student study the **Fractions 1** book prior to studying this book, if he has not already done so.

I have made a set of videos to match many of the lessons in this book. You can access them at https://www.mathmammoth.com/videos/fractions_2.php

This book is meant for fifth grade, and deals in-depth with the following topics:

- simplifying fractions; including simplifying before multiplying;
- multiplication of fractions (and of mixed numbers);
- division of fractions.

We start out by simplifying fractions. Since this is the opposite of making equivalent fractions, studied in *Math Mammoth Fractions 1*, it should be easy for students to understand. We also use the same visual model, just backwards: This time the pie pieces are joined together instead of split apart.

Next comes multiplying a fraction by a whole number. Since this can be solved by repeated addition, it is not a difficult concept.

Multiplying a fraction by a fraction is first explained as taking a certain part of a fraction, in order to teach the concept. After that, students are shown the usual shortcut for the multiplication of fractions.

Then, we find the area of a rectangle with fractional side lengths, and show that the area can be found by multiplying the side lengths. Students multiply fractional side lengths to find areas of rectangles, and represent fraction products as rectangular areas.

Simplifying before multiplying is a process that is not absolutely necessary for fifth graders. I have included it here because it prepares students for the same process in future algebra studies and because it makes fraction multiplication easier.

Students also multiply mixed numbers, and study how multiplication can be seen as resizing or scaling. This means, for example, that the multiplication $(2/3) \times 18$ km can be thought of as finding two-thirds of 18 km.

Next, we study the division of fractions. The first lesson on the topic shows how fractions can be seen *as* divisions; in other words, for example, 5/3 is the same as $5 \div 3$. This of course gives us a means of dividing whole numbers and getting fractional answers (for example, $20 \div 6 = 3\ 2/6$).

The following two lessons focus on two ways to think about division: equal sharing, and fitting the divisor (measurement division), and how they apply to fraction division. Here is an example of equal sharing: If two people equally share 4/5 of a pizza, how much will each person get? This is represented by the division $(4/5) \div 2 = 2/5$.

The last lesson on the topic explores the thought of solving division problems by multiplying, and then introduces reciprocal numbers and the common shortcut for fraction division.

I wish you success in teaching math!

Maria Miller, the author

Helpful Resources and Games on the Internet

We have compiled a list of external Internet resources that match the topics in this book. This list of links includes web pages that offer:

- **online practice** for concepts;

- online **games**, or occasionally, printable games;

- **animations** and interactive **illustrations** of math concepts;

- **articles** that teach a math concept.

We heartily recommend you take a look at the list. Many of our customers love using these resources to supplement the bookwork. You can use the resources as you see fit for extra practice, to illustrate a concept better, and even just for some fun. Enjoy!

https://l.mathmammoth.com/blue/fractions2

SCAN ME

Simplifying Fractions 1

You have learned how to convert a fraction into an equivalent fraction:

Each slice is **split two ways**.

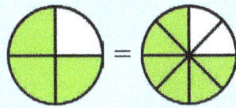

$$\frac{3}{4} = \frac{6}{8}$$

$\times 2$

$\times 2$

What happens if we *reverse* the process?

Then it is called **SIMPLIFYING** or **REDUCING** a fraction:

Every two slices are **joined together**.

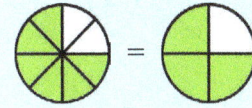

$$\frac{6}{8} = \frac{3}{4}$$

$\div 2$

$\div 2$

1. Simplify the following fractions, filling in the missing parts.

a. Every __*three*__ slices are joined together.

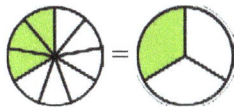

$$\frac{3}{9} = \frac{}{}$$

$\div\ 3$

$\div$

b. Every _____ slices are joined together.

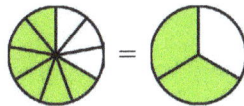

$\div$

$\div$

c. Every _____ slices are joined together.

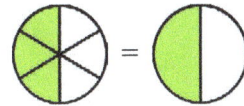

$\div$

$\div$

d. Every _____ slices were joined together.

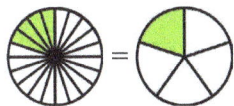

$\div$

$\div$

e. Every _____ parts were joined together.

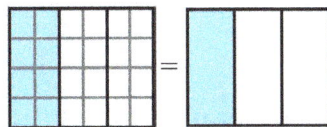

$\div$

$\div$

f. Every _____ parts were joined together.

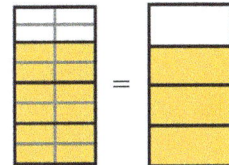

$\div$

$\div$

7

2. Write the simplifying process. You can write the arrows and the divisions to help you.

a. Every _____ slices were joined together.	**b.** Every _____ slices were joined together.	**c.** Every _____ slices were joined together.	**d.** Every _____ slices were joined together.

3. Draw a picture and reduce the fractions.

a. Join together every six parts.	**b.** Join together every four parts.	**c.** Join together every three parts.

4. Reduce the fractions.

a. $\dfrac{6}{16} =$ 　　　　**b.** $\dfrac{15}{25} =$ 　　　　**c.** $\dfrac{28}{32} =$ 　　　　**d.** $\dfrac{12}{42} =$ 　　　　**e.** $\dfrac{18}{27} =$

5. **a.** What happens to the *value* of the fraction in simplification?

b. Why do you think it's called *simplifying* or *reducing* a fraction?

6. Could there ever be a fraction that cannot be simplified? (Why or why not?) Give examples.

In simplification, we need to *divide* the numerator and the denominator by some number, which means the numerator and the denominator both have to be **divisible by this number**—a number that is a **common factor** to both.

Example 1. Since both 28 and 40 are divisible by 4, we can divide the numerator and denominator by four. (This means that each four slices are joined together.)

$$\div \boxed{4}$$
$$\frac{28}{40} = \frac{7}{10}$$
$$\div \boxed{4}$$

Example 2. We cannot find any number that would go into 6 *and* 17 (except 1). So 6/17 is already as simplified as it can be. It is already in its **lowest terms**.

$$\div \boxed{1}$$
$$\frac{6}{17} = \frac{6}{17}$$
$$\div \boxed{1}$$

7. Simplify.

a. $\dfrac{12}{20} =$	**b.** $\dfrac{24}{32} =$	**c.** $\dfrac{3}{15} =$	**d.** $\dfrac{15}{18} =$	**e.** $\dfrac{16}{20} =$

8. Simplify the fractional parts of these mixed numbers. The whole number does not change.

a. $1\dfrac{4}{16} =$	**b.** $5\dfrac{3}{27} =$	**c.** $7\dfrac{5}{20} =$	**d.** $3\dfrac{14}{49} =$

9. You cannot simplify some of these fractions because they are already in lowest terms. Cross out the ones that are already in lowest terms, and simplify the rest.

a. $\dfrac{2}{3}$	**b.** $3\dfrac{4}{28}$	**c.** $\dfrac{6}{13}$	**d.** $\dfrac{6}{33}$
e. $\dfrac{11}{22}$	**f.** $1\dfrac{4}{7}$	**g.** $\dfrac{5}{11}$	**h.** $\dfrac{9}{21}$

10. Tommy is on the track team. He spends 10 minutes warming up before practice, and 10 minutes stretching after practice. All together, he spends a total of one hour for the warm-up, the practice, and the stretching.

What part of the total time is the warm-up time?
Give your answer in lowest terms.

What part of the total time is the actual practice time?
Give your answer in lowest terms.

Simplifying Fractions 2

We can simplify 36/48 in two steps:

$$\overset{\div 6}{\overset{\frown}{\underset{\div 6}{\frac{36}{48}}}} = \overset{\div 2}{\overset{\frown}{\underset{\div 2}{\frac{6}{8}}}} = \frac{3}{4}$$

...or in one step:

$$\overset{\div 12}{\overset{\frown}{\underset{\div 12}{\frac{36}{48}}}} = \frac{3}{4}$$

Both ways are correct.

Since $\frac{3}{4}$ cannot be reduced any further, it is in its **lowest terms**.

Whether you simplify a fraction in several steps or in one step, you get the same result.

When you cannot simplify a fraction any further, the fraction is in its *lowest terms*.

1. Simplify in two steps as indicated. Fill in the missing parts.

a.

$$\overset{\div 10}{\overset{\frown}{\underset{\div 10}{\frac{40}{120}}}} = \frac{\boxed{}}{\boxed{}} \overset{\div 4}{\underset{\div 4}{=}} \frac{\boxed{}}{\boxed{}}$$

You could simplify in one step if you divided by _____.

b.

$$\overset{\div 5}{\overset{\frown}{\underset{\div 5}{\frac{75}{105}}}} = \frac{\boxed{}}{\boxed{}} \overset{\div 3}{\underset{\div 3}{=}} \frac{\boxed{}}{\boxed{}}$$

You could simplify in one step if you divided by _____.

c.

$$\overset{\div 2}{\overset{\frown}{\underset{\div 2}{\frac{42}{98}}}} = \frac{\boxed{}}{\boxed{}} \overset{\div 7}{\underset{\div 7}{=}} \frac{\boxed{}}{\boxed{}}$$

You could simplify in one step if you divided by _____.

2. Simplify the fractions and the fractional parts of the mixed numbers.

a. $\dfrac{60}{200} =$	**b.** $\dfrac{24}{64} =$	**c.** $\dfrac{25}{70} =$
d. $5\dfrac{66}{88} =$	**e.** $3\dfrac{16}{56} =$	**f.** $7\dfrac{36}{60} =$

3. Simplify the improper fractions using the same process as for proper fractions. Lastly, rewrite the simplified fraction as a mixed number.

a.	**b.**	**c.**
$\dfrac{54}{36} = \dfrac{\boxed{}}{\boxed{}} =$	$\dfrac{64}{48} =$	$\dfrac{56}{49} =$

When solving problems with fractions, you should give your answer:

- in lowest terms, and

- as a mixed number, if applicable.

It doesn't matter which you do first (convert into a mixed number or simplify), but usually it is easier to simplify first, then convert the fraction into a mixed number.

4. Simplify the fractions that can be simplified. Give your answer as a mixed number when possible.

a. $\dfrac{14}{29}$	b. $\dfrac{22}{8}$	c. $2\dfrac{8}{15}$
d. $\dfrac{44}{10}$	e. $1\dfrac{21}{35}$	f. $\dfrac{27}{11}$

5. Add or subtract. Give your answer as a mixed number and in lowest terms.

a. $\dfrac{3}{2} + \dfrac{6}{7}$	b. $\dfrac{15}{8} - \dfrac{3}{10}$	c. $5\dfrac{5}{9} + 3\dfrac{7}{12}$

6. Use lines to connect the fractions and mixed numbers to other equivalent ones in the neighboring columns. Try to find as many connections as you can!

$2\dfrac{6}{24}$	$\dfrac{28}{12}$	$1\dfrac{3}{4}$	$2\dfrac{4}{12}$
$\dfrac{14}{8}$	$2\dfrac{2}{8}$	$2\dfrac{5}{15}$	$\dfrac{21}{12}$
$2\dfrac{1}{3}$	$\dfrac{7}{4}$	$\dfrac{9}{4}$	$2\dfrac{1}{4}$

7. Three students did a problem that said, "Simplify to lowest terms."

$\dfrac{96}{120} = \dfrac{12}{} = \dfrac{}{}$ **a.** Jerry divided the numerator and the denominator by 8, and then by 3.	$\dfrac{96}{120} = \dfrac{8}{}$ **b.** Mark divided them by 12.	$\dfrac{96}{120} = \dfrac{}{}$ **c.** Nancy divided them by 24.

Who got it right? _____ Who didn't? _____

Why? _____

8. A computer screen is 1600 pixels wide. A horizontal line on the screen is 1200 pixels wide.

   a. What part of the width of the screen does the line take up?

   b. How wide should the line be if you want it to take up exactly 3/8 of the total width of the screen?

9. Simplify. Place the letter from each problem under the correct answer, and solve the riddle.

WHY ARE TEDDY BEARS NEVER HUNGRY?

$$\frac{3}{4}\ \frac{2}{5}\ \frac{1}{2}\ \frac{2}{7}\qquad \frac{1}{4}\ \frac{3}{5}\ \frac{1}{2}\qquad \frac{1}{4}\ \frac{2}{3}\ \frac{1}{6}\ \frac{1}{4}\ \frac{2}{7}\ \frac{1}{3}\qquad \frac{1}{3}\ \frac{3}{4}\ \frac{5}{6}\ \frac{3}{10}\ \frac{3}{10}\ \frac{1}{2}\ \frac{3}{7}$$

Because ☐☐☐☐ ☐☐☐ ☐☐☐☐☐☐ ☐☐☐☐☐☐☐ .

E. $\dfrac{5}{10} =$	**H.** $\dfrac{4}{10} =$	**R.** $\dfrac{6}{10} =$	**E.** $\dfrac{18}{36} =$	**L.** $\dfrac{6}{9} =$
S. $\dfrac{12}{36} =$	**A.** $\dfrac{15}{60} =$	**S.** $\dfrac{24}{72} =$	**W.** $\dfrac{3}{18} =$	**E.** $\dfrac{30}{60} =$
T. $\dfrac{15}{20} =$	**Y.** $\dfrac{4}{14} =$	**Y.** $\dfrac{8}{28} =$	**F.** $\dfrac{15}{50} =$	**D.** $\dfrac{15}{35} =$
A. $\dfrac{20}{80} =$	**F.** $\dfrac{18}{60} =$	**T.** $\dfrac{12}{16} =$	**A.** $\dfrac{25}{100} =$	**U.** $\dfrac{25}{30} =$

Multiply Fractions and Whole Numbers 1

1. Write a multiplication sentence to match the illustration.

a. Three copies of 4/5:	**b.** Three groups of 7/9:	**c.** Two groups of 7/8:
How many *fifths* in total?	How many *ninths* in total?	How many *eighths* in total?
$\underline{\ \ 3\ \ } \times \dfrac{4}{5} = \dfrac{\ \ }{\ \ }$	$\underline{\ \ \ \ } \times \dfrac{\ \ }{\ \ } = \dfrac{\ \ }{\ \ }$	$\underline{\ \ \ \ } \times \dfrac{\ \ }{\ \ } = \dfrac{\ \ }{\ \ }$

One way to view a product of a whole number and a fraction is to think of so many copies (groups) of the fraction, and to find the total number of pieces.

In this multiplication, the denominator does not change (we still have the same *kind* of parts).

Example 1. $8 \times \dfrac{3}{4}$ means 8×3 pieces, or 24 pieces. Each piece is a fourth. So, we get $\dfrac{24}{4}$.

We don't leave the answer as an improper fraction, so we write $\dfrac{24}{4}$ as 6.

---

2. Multiply. Remember to give your final answer as a <u>mixed number</u>. The pie pictures can help.

a. $3 \times \dfrac{7}{10} =$	**b.** $4 \times \dfrac{7}{9} =$	**c.** $3 \times \dfrac{5}{8} =$

3. Solve. Give your answer in lowest terms (simplified) and as a mixed number. Study the example.

a. $6 \times \dfrac{4}{9} = \dfrac{24}{9} = \dfrac{8}{3} = 2\dfrac{2}{3}$	**b.** $4 \times \dfrac{7}{10} =$
c. $2 \times \dfrac{11}{20} =$	**d.** $9 \times \dfrac{2}{15} =$

4. Erica has beverage glasses that hold 3/8 liters each.
   How much water does she need to fill four of them?

Example 2. Multiplication can be done in either order. (Multiplication is *commutative*.)

So, $\dfrac{3}{10} \times 5$ is equal to $5 \times \dfrac{3}{10}$. They both equal $\dfrac{5 \times 3}{10} = \dfrac{15}{10}$. This simplifies to $\dfrac{3}{2}$, which is $1\dfrac{1}{2}$.

5. Solve. Give your answer in lowest terms (simplified) and as a mixed number.

a. $\dfrac{15}{6} \times 2 =$	**b.** $6 \times \dfrac{7}{100} =$
c. $\dfrac{1}{12} \times 16 =$	**d.** $2 \times \dfrac{35}{100} =$
e. $\dfrac{9}{20} \times 10 =$	**f.** $\dfrac{7}{15} \times 7 =$

6. Marlene wants to triple this recipe (make three times as much). How much of each ingredient will she need?

> **Brownies**
>
> 3/4 cup butter
> 1 1/2 cups brown sugar
> 4 eggs
> 1 1/4 cups cocoa powder
> 1/2 cup flour
> 2 tsp vanilla

7. William asked 20 fifth graders how much time they spent on housework and chores the day before. He then rounded the answers to the nearest 1/8 hour. The line plot shows his results. Each x-mark corresponds to one fifth grader.

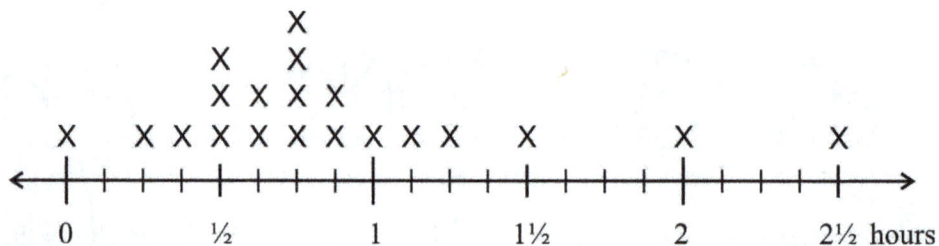

```
                              X
                    X      X
                    X  X  X
     X       X  X  X  X  X  X  X  X  X    X        X        X
  <--+--+--+--+--+--+--+--+--+--+--+--+--+--+--+--+--+--+--+--+-->
     0       ½         1        1½        2        2½ hours
```

a. Exclude the three students who did the least housework and three who did the most, and fill in:

Most students used between _____ and _____ hours for housework and chores.

b. How many students spent 45 minutes on housework and chores?

c. The average for this data is 7/8 hours. Use this to calculate how many hours these 20 fifth graders used for housework in total.

Multiply Fractions and Whole Numbers 2

We can calculate a fractional part of a quantity or number by <u>using multiplication</u>.

Essentially, **the word "of" translates into multiplication**.

Example 1. $\frac{3}{10}$ of 120

$$\downarrow$$

$$\frac{3}{10} \times 120$$

Here is a connection. You have previously learned how to find a fractional part of a quantity using division first. For example, you have learned to find 2/5 of $40 this way:

- First, divide $40 by 5 to find 1/5 of it. It is $8.

- Then, multiply that by 2 to get $16.

Notice, this is the calculation $\frac{\$40}{5} \times 2$.

See Example 2 on the right. We get the same answer with fraction multiplication!

Both methods are essentially the same: you divide by 5 and multiply by 2, it is just done in two different orders.

Example 2.

$$\frac{2}{5} \text{ of } \$40$$

$$\downarrow$$

$$= \frac{2}{5} \times \$40$$

$$= \frac{2 \times \$40}{5} = \frac{\$80}{5} = \$16$$

To recap, all of these expressions have the same value:

$$\frac{3}{10} \times \$120 \qquad \frac{\$120 \times 3}{10} \qquad \frac{\$120}{10} \times 3 \qquad \$120 \times \frac{3}{10}$$

(...and a few others, too, that we'd get by switching the order of the multiplication.)

Notice, all of them have a division by 10, and a multiplication by 3.

1. Write a multiplication equation.

a. Find $\frac{1}{2}$ of 60. $\frac{1}{2} \times 60 = $ _____	**b.** Find $\frac{1}{3}$ of 150. $\frac{}{} \times$ _____ $=$ _____	**c.** Find $\frac{1}{8}$ of 24. $\frac{}{} \times$ _____ $=$ _____
d. Find $\frac{4}{5}$ of 100. $\frac{}{} \times$ _____ $=$	**e.** Find $\frac{2}{3}$ of 36. $\frac{}{} \times$ _____ $=$	**f.** Find $\frac{5}{6}$ of 30. $\frac{}{} \times$ _____ $=$

2. Compare the problems in each box. How can you use the result from the top problem to help you solve the bottom one?

a.	b.	c.
$\dfrac{1}{4} \times 60 =$ _____	$\dfrac{1}{5} \times 45 =$ _____	$\dfrac{1}{9} \times 180 =$ _____
$\dfrac{3}{4} \times 60 =$ _____	$\dfrac{2}{5} \times 45 =$ _____	$\dfrac{8}{9} \times 180 =$ _____

3. Calculate. Note the units, and include them in your answer, also.

a. $\dfrac{3}{8} \times 40$ km =	**b.** $\dfrac{2}{5} \times \$60 =$
c. $\dfrac{7}{10} \times 500$ kg =	**d.** $\dfrac{3}{4} \times 32$ in =

4. Find the following quantities.

a. 2/5 of 35 lb

b. 4/9 of 180 km

5. Make up a real-life situation where you can you use the calculation $\dfrac{2}{3} \times 600$.

6. Dad is building a shelf that is 4 meters long. He wants to use 2/5 of it for gardening supplies and the rest for tools. How long is the section of the shelf that is for gardening supplies? (*Hint: Use centimeters.*)

7. **a.** Janet and Sandy earned $81 for doing yard work. They divided the money unequally so that Janet got 2/3 of it and Sandy got the rest. How much money did each girl get?

b. What happens if the amount they earned is $80 instead?

Multiply Fractions by Fractions 1

We have studied how to find a fractional part of a whole number using multiplication.

For example, $\frac{3}{5}$ of 80 is written as the multiplication $\frac{3}{5} \times 80$.

REMINDER: The word "of" in this context translates into multiplication.

Now let's examine how we can use the same idea to find **a fractional part of a fraction.**

1. <u>First</u> find a fractional part of the given fraction **visually**. You can think of a leftover pizza piece, which you are sharing equally with some other people. Then write a multiplication.

a. $\frac{1}{2}$ of [circle] is [circle]

$\frac{1}{2} \times \frac{1}{2} =$

b. $\frac{1}{2}$ of [circle] is [circle]

$\frac{}{} \times \frac{}{} = \frac{}{}$

c. $\frac{1}{2}$ of [circle] is [circle]

$\frac{}{} \times \frac{}{} = \frac{}{}$

d. $\frac{1}{3}$ of [circle] is [circle]

$\frac{}{} \times \frac{}{} = \frac{}{}$

e. $\frac{1}{3}$ of [circle] is [circle]

$\frac{}{} \times \frac{}{} = \frac{}{}$

f. $\frac{1}{3}$ of [circle] is [circle]

$\frac{}{} \times \frac{}{} = \frac{}{}$

g. $\frac{1}{4}$ of [circle] is [circle]

$\frac{}{} \times \frac{}{} = \frac{}{}$

h. $\frac{1}{4}$ of [circle] is [circle]

$\frac{}{} \times \frac{}{} = \frac{}{}$

i. $\frac{1}{4}$ of [circle] is [circle]

$\frac{}{} \times \frac{}{} = \frac{}{}$

2. Did you notice a shortcut? If so, write it here. Use examples, such as (1/5) × (1/2) and (1/4) × (1/6).

Shortcut for multiplying a *unit fraction* by another *unit fraction*:
(A unit fraction is of the form 1/*n* where *n* is a whole number.)

<table>
<tr><td colspan="2" align="center">**Shortcut: multiplying unit fractions**</td></tr>
<tr><td>To multiply fractions of the form 1/n where n is a whole number, simply multiply the denominators to get the new denominator.</td><td>**Example 1.** $\dfrac{1}{4} \times \dfrac{1}{5} = \dfrac{1}{20}$ and $\dfrac{1}{2} \times \dfrac{1}{6} = \dfrac{1}{12}$</td></tr>
</table>

3. Multiply.

a. $\dfrac{1}{9} \times \dfrac{1}{2}$	b. $\dfrac{1}{13} \times \dfrac{1}{3}$	c. $\dfrac{1}{5} \times \dfrac{1}{20}$

What about finding some other kind of fractional part? Let's again compare this to whole numbers.

Review: To find $\dfrac{3}{4}$ of 16, or in other words $\dfrac{3}{4} \times 16$, you can first find $\dfrac{1}{4}$ of 16, which is 4.

Then just take that three times, which is 12. In other words, $\dfrac{3}{4} \times 16$ becomes $3 \times (\frac{1}{4}$ of 16$) = 12$.

We can use the same idea when finding a fractional part of another fraction.

4. Color in the answer. Compare the problems in each box.

5. Two-thirds of a pizza is left from last night. You eat half of what is left.

   a. Write a fraction multiplication to match this situation.
   You can also draw a picture to help you.

   b. Is your final answer a fraction of the *original* pizza, or of the portion that was left over?

Example 2. To find $\frac{4}{5}$ of $\frac{1}{7}$, first find $\frac{1}{5}$ of $\frac{1}{7} = \frac{1}{35}$, and then take that four times to get $\frac{4}{35}$.

Multiplying a fraction by a fraction means taking that fractional part *of* the fraction.
It is just like taking a certain part of the leftovers, when what is left over is a fraction.

6. Solve. You can find the answer to the bottom problem based on the top problem in each box.

a. $\frac{1}{5} \times \frac{1}{7} =$	**b.** $\frac{1}{6} \times \frac{1}{4} =$	**c.** $\frac{1}{8} \times \frac{1}{3} =$
$\frac{2}{5} \times \frac{1}{7} =$	$\frac{5}{6} \times \frac{1}{4} =$	$\frac{3}{8} \times \frac{1}{3} =$

What about generic fraction multiplication problems? For example, how can we do $\frac{5}{8} \times \frac{6}{7}$?

Mathematically, we can treat this as $5 \times \frac{1}{8} \times 6 \times \frac{1}{7}$, and then change the order of the factors to

get $5 \times 6 \times \frac{1}{8} \times \frac{1}{7}$, which is equal to $5 \times 6 \times \frac{1}{56} = \frac{30}{56}$.

Essentially, the numerators get multiplied, and the denominators get multiplied.

A shortcut for fraction multiplication: Multiply the numerators to get the numerator for the product.
Multiply the denominators to get the denominator for the product.

Example 3. Give your final answer simplified and as a mixed number.	**Example 4.** Notice how we can write the whole number 5 as 5/1:
$\frac{4}{5} \times \frac{11}{8} = \frac{4 \times 11}{5 \times 8} = \frac{44}{40} = \frac{11}{10} = 1\frac{1}{10}$	$\frac{3}{7} \times 5 = \frac{3}{7} \times \frac{5}{1} = \frac{3 \times 5}{7 \times 1} = \frac{15}{7} = 2\frac{1}{7}$

7. Multiply. Give your answers in the lowest terms and as mixed numbers, if possible.

a. $\frac{3}{9} \times \frac{2}{9}$	**b.** $\frac{11}{12} \times \frac{1}{6}$
c. $\frac{1}{3} \times \frac{3}{13}$	**d.** $9 \times \frac{2}{3}$
e. $\frac{2}{9} \times \frac{6}{7}$	**f.** $10 \times \frac{5}{7}$

Multiply Fractions by Fractions 2

Example 1. Henry finished 1/4 of a job he was given in one day. The next day, he finished half of what was left. Now what part of the task is left to do?

After the first day of work, he has 3/4 of the job left:

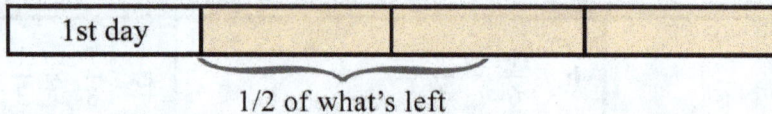

1st day			

1/2 of what's left

Then he finished *half* of that. This means we need to figure out half of 3/4. This is found with fraction multiplication: $\dfrac{1}{2} \times \dfrac{3}{4} = \dfrac{3}{8}$. What does this 3/8 signify?

It is 3/8 of the original whole:

1st day						

1. There was 1/4 of the pizza left. Marie ate 2/3 of that.

   a. Write a fraction multiplication. You can also draw a picture.

   b. What part of the *whole* (original) pizza did she eat?

   c. What part of the *whole* (original) pizza is left now?

2. Theresa has painted 5/8 of the room.

   a. What part is still left to paint?

   b. Now, Theresa has painted *half* of what was still left.

   Write a fraction multiplication.

   Use the bar model on the right to help you.
   What part of the room is still left to paint?

3. Sally wants to make 1/3 of
   the recipe on the right.
   How much does she need
   of each ingredient?

Carob Brownies

3 cups sweetened carob chips
8 tablespoons olive oil
2 eggs
1/2 cup honey
1 teaspoon vanilla
3/4 cup whole wheat flour
3/4 teaspoon baking powder
1 cup walnuts or other nuts

4. Multiply. Give your answers in the lowest terms (simplified) and as mixed numbers, if possible.

a. $\dfrac{3}{4} \times \dfrac{7}{8} =$	**b.** $\dfrac{7}{10} \times \dfrac{8}{5} =$
c. $\dfrac{9}{20} \times \dfrac{4}{5} =$	**d.** $\dfrac{2}{5} \times 18 =$
e. $30 \times \dfrac{5}{7} =$	**f.** $\dfrac{9}{4} \times \dfrac{8}{11} =$

5. Ted has completed 2/3 of a job that his boss gave him.

   a. What part is still left to do?

   b. Now Ted completes a third of what was still left to do.
   Draw a bar model and write a fraction multiplication.
   What part of the original job is still left undone?

   What part is completed?

6. For an upcoming get-together, Alison uses the recipe on the right.

   a. Let's say that each guest drinks one serving of coffee.
   Find the amount of <u>ground coffee</u> she will need for 30 guests.

Coffee (5 servings)

3 1/2 cups water
1/4 cup ground coffee

   b. Now let's say that each guest drinks two servings, and that she
   will have 50 guests. Find the amount of ground coffee she will need.

Puzzle Corner	**a.** $\dfrac{}{} \times \dfrac{6}{7} = \dfrac{1}{7}$	**b.** $\dfrac{}{} \times \dfrac{1}{4} = \dfrac{5}{16}$	**c.** $\dfrac{}{} \times \dfrac{2}{5} = \dfrac{3}{10}$
Find the missing factors.			

21

Fraction Multiplication and Area

What is the area of this rectangle?

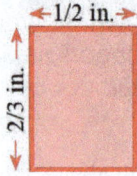

Notice, its side lengths are *fractional* (1/2 inch and 2/3 inch).

Let's extend its sides and draw a square inch around it.

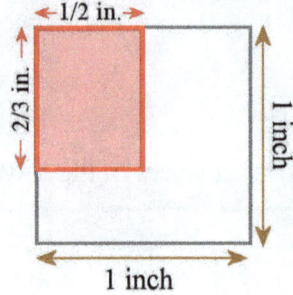

Surely the area of our rectangle is less than a half square inch. But how much is the area exactly?

To solve this problem, let's draw a grid inside our square inch:

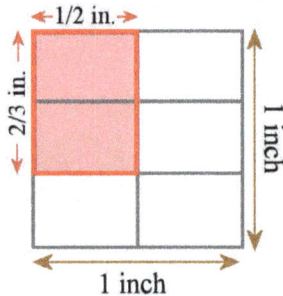

Now it is easy to see that the area of the colored rectangle is exactly 2/6 or 1/3 of the square inch.

(Why? Because the square inch is divided into 6 equal parts, and our rectangle covers two of them).

Notice that we get the same result (1/3 square inch) if we *multiply* the side lengths, using fraction multiplication:

$$\frac{2}{3} \text{ in} \times \frac{1}{2} \text{ in} = \frac{2}{6} \text{ in}^2 = \frac{1}{3} \text{ in}^2$$

1. Each picture shows some kind of square unit, and a colored rectangle. Figure out the side lengths and the area of the rectangle from the picture.

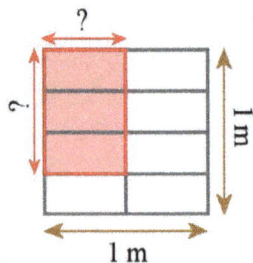

a.

Side lengths: ⬜—— m and ⬜—— m

Area (from the picture): ⬜—— m²

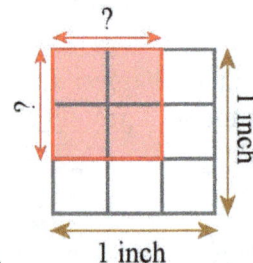

b.

Side lengths: ⬜—— in and ⬜—— in

Area (from the picture): ⬜—— in²

2. Again, figure out the side lengths of the colored rectangle from the picture. Then multiply the side lengths to find its area. <u>Check that the area you get by multiplying is the same as what you can see from the picture.</u>

a.

? ? 1 m 1 m

Side lengths: ____ m and ____ m

Area (by multiplication):

____ m × ____ m =

b.

1 inch 1 inch

Side lengths: ____ in and ____ in

Area (by multiplication):

____ in × ____ in =

c.

1 m 1 m

Side lengths: ____ m and ____ m

Area (by multiplication):

____ m × ____ m =

d.

1 km 1 km

Side lengths: ____ km and ____ km

Area (by multiplication):

____ km × ____ km =

3. Shade a rectangle inside the square so that its area can be found by the fraction multiplication.

a. $\dfrac{1}{4}$ m × $\dfrac{1}{2}$ m = $\dfrac{1}{8}$ m^2

b. $\dfrac{1}{2}$ in × $\dfrac{4}{6}$ in = $\dfrac{4}{12}$ in^2

c. $\dfrac{3}{4}$ ft × $\dfrac{2}{7}$ ft =

d. $\dfrac{3}{5}$ km × $\dfrac{5}{6}$ km =

23

The area of this rectangle *can* be found by multiplication:

$\frac{3}{4}$ m × $\frac{1}{3}$ m = $\frac{1}{4}$ m²; however, we want to <u>verify</u> this using a visual method.

For that reason, let's sketch a unit square around the rectangle and tile it.

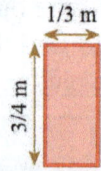

We need to extend the sides of the rectangle to draw the square. The 1/3-meter side simply needs to be three times as long to make it 1 meter.

Then, divide the side that is 3/4 meters long into three equal parts—each part is 1/4 m long. Then extend that side by another 1/4 meter.

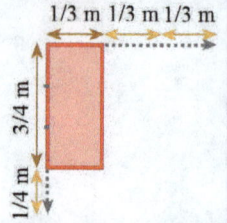

Lastly, draw the entire square. Draw gridlines to show the tiles within the square meter: one side is divided into 3 equal parts, and the other into 4 equal parts. We get 12 tiles.

Now it is easy to see that the area of the colored rectangle is 3 tiles out of 12, or 3/12 of a square meter. That simplifies to 1/4 of a square meter.

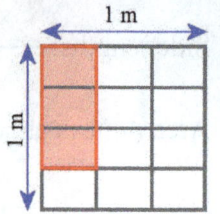

4. Extend the sides of the rectangle so you get a square meter (unit square). Draw gridlines into the square as in the example above. Write a multiplication for the area of the colored rectangle. <u>Verify</u> that the area you get by multiplying is the same as what you can see in the picture.

a.   1/3 m

1/3 m

Area: 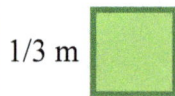 m × 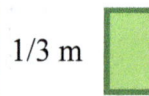 m =

b.   1/5 m

1/3 m

c.   1/5 m

1/2 m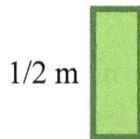

d.   1/4 m

1/4 m

5. Extend the sides of the rectangle so you get a square meter (unit square). Draw gridlines into the square as in the example above. Write a multiplication for the area of the colored rectangle. <u>Verify that the area you get by multiplying is the same as what you can see in the picture.</u>

a. 3/4 m 1/2 m 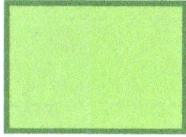	**b.** 2/5 m 3/4 m 
c. 2/3 m 2/3 m 	**d.** 3/5 m 1/2 m 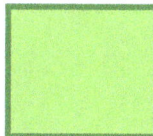
e. 3/4 m 3/4 m 	**f.** 5/6 m 1/2 m 

6. In the pictures below, the outer square is <u>one square unit</u>. Write a multiplication for the area of the colored rectangle. This time, we are not using meters or inches, just "units" and "square units," and you do not have to include those in the multiplication (simply write the fractions without any units).

a. 	___ × ___ =	**b.** 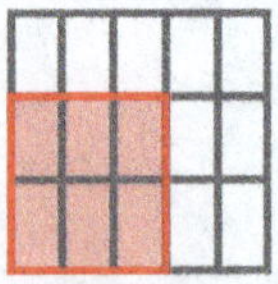	___ × ___ =
c. 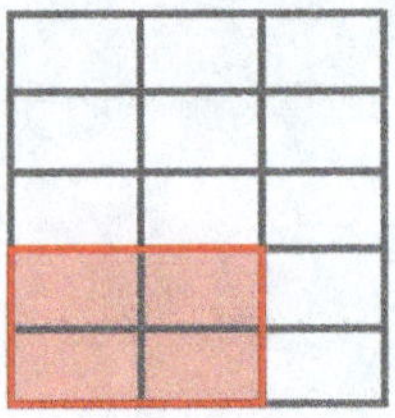	___ × ___ =	**d.** 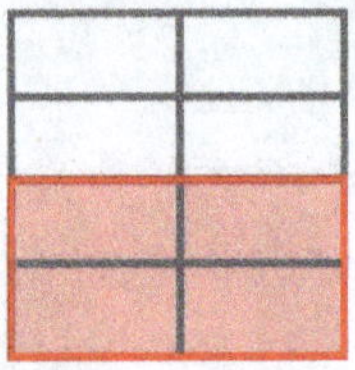	___ × ___ =

7. **a.** In the space on the right, sketch a 1 inch by 1 inch square
   (each side of the square measures 1 inch).
   Use a ruler to measure the sides.
   What is its area?

   b. Inside your square, draw now a rectangle with 3/4 in
   and 5/8 in sides so that the two sides of the rectangle
   touch the sides of the square. See the illustration below
   (not to scale).

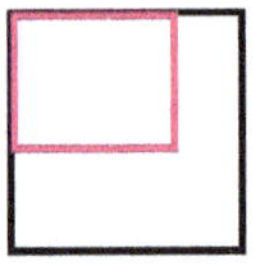

   c. Find the area of your rectangle. Is it more or less than half a square inch?

8. Find the area of a rectangular suburb that is 3 km by 500 m, in square kilometers.

9. A rectangle has 3/10 cm and 7/10 cm sides. Calculate the area of the rectangle in square centimeters using both fractions and decimals (calculate it two times):

    a. Using fractions:

    b. Using decimals:

10. **a.** A stamp measures 7/8 in by 3/4 in. Amanda puts six of them onto an envelope, side by side. Find the total area these stamps cover.

    b. The envelope is 8 in by 5 in. *About* what part of the envelope do the six stamps cover?

Puzzle Corner

Which has a larger area, a square with 7/8-mile sides, or a rectangle that is 1/4 mile by 3 miles?
How much larger?

Simplifying Before Multiplying

(This lesson is optional.)

We will start using a **new notation** to indicate simplifying fractions. When a numerator or a denominator gets simplified, we will cross it out with a slash, and write the *new* numerator or denominator next to it (above or below).

$$\frac{\overset{3}{\cancel{12}}}{\underset{5}{\cancel{20}}} = \frac{3}{5}$$

The number you divide by (the 4) does *not* get indicated in any way! You only think about it: "I divide 12 by 4, and get 3. I divide 20 by 4, and get 5."

You may not see any advantage over the "old" method yet, but this shortcut will come in handy soon.

$$\frac{\overset{7}{\cancel{35}}}{\underset{11}{\cancel{55}}} = \frac{7}{11}$$

1. Simplify, using the new notation.

a. $\dfrac{14}{16}$	**b.** $\dfrac{33}{27}$	**c.** $\dfrac{12}{26}$	**d.** $\dfrac{9}{33}$

Before multiplying, we can write an equivalent, simpler fraction in the place of a fraction.

In the example on the right, 4/10 is simplified to 2/5 before multiplying. We write a tiny "2" above the "4" and a tiny "5" below the "10".

Why does this work? We can write 2/5 in the place of 4/10, since they are *equivalent*. The value does not change.

Example 1.

$$\frac{3}{7} \times \frac{\overset{2}{\cancel{4}}}{\underset{5}{\cancel{10}}} = \frac{6}{35}$$

2. Simplify one or both fractions before multiplying. Use equivalent fractions. Look at the example.

a. $\dfrac{\overset{3}{\cancel{6}}}{\underset{5}{\cancel{10}}} \times \dfrac{\overset{1}{\cancel{2}}}{\underset{7}{\cancel{14}}} = \dfrac{3 \times 1}{5 \times 7} = \dfrac{3}{35}$	**b.** $\dfrac{2}{4} \times \dfrac{3}{15} =$
c. $\dfrac{8}{12} \times \dfrac{1}{2} =$	**d.** $\dfrac{8}{32} \times \dfrac{14}{21} =$
e. $\dfrac{6}{15} \times \dfrac{6}{9} =$	**f.** $\dfrac{27}{45} \times \dfrac{21}{49} =$

You can also simplify "crisscross." Look at the example on the right: →

We simplify 3 and 6, writing 1 and 2 in their place. Think of it as the fraction 3/6 being simplified into 1/2, but the 3 and 6 are across from each other.

$$\frac{7}{\overset{}{\underset{2}{6}}} \times \frac{\overset{1}{3}}{9} = \frac{7}{18}$$

Why are we allowed to simplify in such a manner?

Compare the above problem to this one: $\frac{7}{9} \times \frac{3}{6}$. (It is almost the same, isn't it?) Surely you can see that in this problem, we *could* simplify 3/6 to 1/2 before multiplying. And, these two multiplication problems are essentially the *same* problem, because they both lead to the same expression and the same answer:

the first one becomes $\frac{7 \times 3}{6 \times 9} = \frac{21}{54}$, and the second one becomes $\frac{7 \times 3}{9 \times 6} = \frac{21}{54}$ (without simplifying).

Therefore, since you can simplify 3/6 into 1/2 in the one problem, you can do the same in the other also.

3. Simplify "crisscross" before you multiply.

a. $\frac{8}{9} \times \frac{6}{11}$	**b.** $\frac{3}{16} \times \frac{8}{5}$	**c.** $\frac{4}{7} \times \frac{1}{12}$

You can even simplify crisscross several times before multiplying.

But you can only simplify one pair at a time (that is, one numerator TO one denominator).

$$\frac{\overset{1}{3}}{15} \times \frac{5}{\underset{2}{6}}$$

First, simplify 3 and 6 into 1 and 2.

$$\frac{\overset{1}{\cancel{3}}}{\underset{3}{\cancel{15}}} \times \frac{\overset{1}{\cancel{5}}}{\underset{2}{\cancel{6}}} = \frac{1}{6}$$

Then simplify 5 and 15 into 1 and 3.

4. Simplify before you multiply.

a. $\frac{7}{8} \times \frac{2}{7}$	**b.** $\frac{3}{5} \times \frac{5}{6}$	**c.** $\frac{5}{12} \times \frac{4}{10}$
d. $\frac{9}{15} \times \frac{3}{18}$	**e.** $\frac{8}{11} \times \frac{3}{4}$	**f.** $\frac{12}{100} \times \frac{4}{15}$

5. What is the value of $\frac{25}{36} \times 36$?

   Why?

Example 3. We can think of the problem $\frac{27}{45} \times 45$ in two manners:

(1) Think of the fraction line as division. The problem is therefore the same as $27 \div 45 \times 45$.
Whenever you multiply and divide by the same number, they cancel each other out.

(2) We can write 45 as the fraction 45/1. The problem then becomes $\frac{27}{45} \times \frac{45}{1}$.

Now we simplify crisscross, and multiply: $\frac{27}{\cancel{45}} \times \frac{\cancel{45}^{1}}{1} = \frac{27}{1} = 27$.

6. Simplify and multiply. You can write the whole number as a fraction (with a denominator of 1) first.

a. $\frac{82}{77} \times 77 =$	**b.** $13 \times \frac{49}{13} =$	**c.** $\frac{5}{6} \times 24 =$	**d.** $54 \times \frac{2}{9} =$

7. A toy block is 3/8 in tall. How tall is a stack of 8 of them?

A stack of 20 of them?

8. Sandra buys 3/4 kg of meat every week. How much meat does she buy in a year?

9. How does simplifying before multiplying change the final answer of a fraction multiplication problem?

To multiply three or more fractions, the same principles apply. We simply multiply all the numerators and all the denominators. And you can simplify before you multiply.

Puzzle Corner

a. $\frac{3}{5} \times \frac{1}{7} \times \frac{5}{6} =$	**b.** $\frac{7}{12} \times \frac{3}{5} \times \frac{6}{7} =$	**c.** $\frac{1}{12} \times \frac{4}{3} \times \frac{6}{7} =$
d. $\frac{9}{10} \times \frac{5}{2} \times \frac{2}{7} =$	**e.** $\frac{4}{5} \times \frac{9}{8} \times \frac{10}{24} =$	**f.** $\frac{2}{9} \times \frac{6}{7} \times \frac{7}{8} =$

Multiplying Mixed Numbers

Multiplying mixed numbers is not difficult at all.

- First, change the mixed numbers to fractions.
- Then multiply the fractions.
- Give your answer as a mixed number and in lowest terms.

The most difficult part of this is to **remember *not* to multiply the mixed numbers until you have <u>first changed them into fractions</u>.**

$$1\frac{2}{3} \times 2\frac{5}{6}$$

$$\frac{5}{3} \times \frac{17}{6} = \frac{85}{18} = 4\frac{13}{18}$$

Estimation: 1 2/3 × 3 = 5. The answer is fairly close to 5, so it is reasonable.

Optionally, if you know how, it can really help to simplify before multiplying, because then the numerators and the denominators become smaller numbers.

Note: simplify ONLY after you have changed the mixed numbers to fractions, not before.

You can always use estimation to check that your answer is reasonable (not too big or too small).

$$4\frac{2}{9} \times 3\frac{3}{8}$$

$$\frac{\overset{19}{\cancel{38}}}{\underset{1}{\cancel{9}}} \times \frac{\overset{3}{\cancel{27}}}{\underset{4}{\cancel{8}}} = \frac{57}{4} = 14\frac{1}{4}$$

Estimation: 4 × 3 ½ = 14. The answer 14 ¼ is close to that, so it makes sense.

1. Multiply. Don't forget: After you change the mixed numbers into fractions, you can simplify crisscross to make things easier for yourself! Use estimation to **check that your answer is reasonable** (not too big or too small).

a. $2\frac{1}{4} \times 1\frac{1}{2}$ ↓ ↓	**b.** $5\frac{1}{5} \times \frac{1}{6}$
c. $4\frac{1}{2} \times 1\frac{1}{5}$	**d.** $3\frac{1}{3} \times 2\frac{1}{10}$

2. **a.** A carpet is 5 ½ feet wide and 7 ½ feet long.
   How many square feet does it cover?

   b. A room is 12 ft by 20 ft. *About* what part of the
   floor area does the carpet cover? Use estimation
   (rounded numbers).

3. An student solved $2\frac{1}{2} \times 1\frac{1}{2}$ wrongly like this:

   "First, I multiply the whole numbers: 2 × 1 = 2. Then I multiply the fractional parts: $\frac{1}{2} \times \frac{1}{2} = \frac{1}{4}$*.*
   Lastly, I add those to get 2 ¼."

   Study the visual model and the calculations below. Then use the model to explain why the above
   method is wrong.

   Area 1: 2 × 1 = 2 square units
   Area 2: ½ × 1 = ½ square unit
   Area 3: 2 × ½ = 1 square unit
   Area 4: ½ × ½ = ¼ square unit

   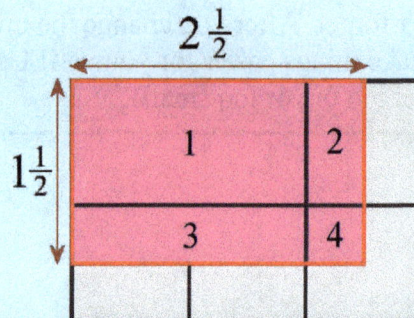

4. Alice is going to make this recipe
   1 ½ times. Calculate the new amount
   of each ingredient for her. Write the
   new amounts on the lines in front of
   the numbers in the recipe.

Cheese Ball
_____ 2 packages cream cheese
_____ 2 ½ cups shredded Cheddar cheese
_____ 1 ½ cups chopped pecans
_____ 1 teaspoon grated onion

5. Practice some more. Change any mixed numbers into fractions before multiplying.

a. $2 \times 7\frac{1}{3}$	**b.** $2\frac{1}{9} \times \frac{1}{3}$
c. $7 \times 2\frac{4}{7}$	**d.** $\frac{7}{8} \times 2\frac{1}{5}$
e. $3\frac{3}{10} \times 2\frac{1}{3}$	**f.** $1\frac{1}{8} \times 2\frac{4}{5}$

6. In the US "letter" size paper measures 8 ½ inches × 11 inches.

   a. What is the area of this kind of paper in square inches?

   b. If you use ½-inch margins on all four sides, what is the real writing area in square inches?

Multiplication as Scaling/Resizing

You know that **scaling** means **expanding or shrinking something** by some factor.

We use **multiplication** to accomplish this. The number we multiply by is called the **scaling factor**.

Example 1. When a stick 40 pixels long is scaled to be 3/5 as long as it was, it will shrink!

$$\text{——} \quad \rightarrow \quad \text{—}$$

We could write this type of a multiplication equation: $(3/5) \times \text{——} = \text{—}$.

Using the length of 40 pixels, we write $(6/10) \times 40 \text{ px} = 24 \text{ px}$ or $0.6 \times 40 \text{ px} = 24 \text{ px}$.

Example 2. The multiplication $(1\,2/3) \times 18$ km means taking the distance of 18 km one and two-thirds times. We're scaling the quantity 18 km by the factor 1 2/3.

To calculate it, we can multiply in parts: take 1×18 km, and $(2/3) \times 18$ km, and add those. Since two-thirds of 18 km is 12 km, then **$(1\,2/3) \times 18$ km is 18 km + 12 km = 30 km.**

1. The stick and other quantities are being scaled—either expanded or shrunk. Find the quantity after scaling. Compare the problems in each box.

a.	b.	c.
$\frac{1}{2} \times \text{——} = \text{—}$	$\frac{1}{4} \times \text{——} = \text{–}$	$\frac{5}{8} \times 400 \text{ km} = \underline{\hspace{2cm}}$
$\frac{1}{2} \times 50 \text{ px} = \underline{\hspace{1.5cm}} \text{ px}$	$\frac{1}{4} \times 40 \text{ px} = \underline{\hspace{1.5cm}} \text{ px}$	$2\frac{5}{8} \times 400 \text{ km} = \underline{\hspace{2cm}}$
$1\frac{1}{2} \times \text{——} = \text{———}$	$2\frac{1}{4} \times \text{——} = \text{————}$	**d.**
$1\frac{1}{2} \times 50 \text{ px} = \underline{\hspace{1.5cm}} \text{ px}$	$2\frac{1}{4} \times 40 \text{ px} = \underline{\hspace{1.5cm}} \text{ px}$	$\frac{3}{5} \times \$600 = \underline{\hspace{2cm}}$
		$3\frac{3}{5} \times \$600 = \underline{\hspace{2cm}}$

2. A 1200×800 photo (in pixels) is scaled by scaling factor s.

   a. If you want the resulting photo to be slightly smaller than the original, what kind of number would you use for s?

   b. If $s = 2\,\frac{3}{4}$, calculate the dimensions of the resulting photo.

3. Will the resulting stick be longer or shorter than the original—or equally long? You do not have to calculate anything. Compare.

a. $\dfrac{9}{8}$ × —— is longer/shorter than ——.	**b.** $\dfrac{3}{7}$ × —— is longer/shorter than ——.
c. $3\dfrac{2}{100}$ × —— is longer/shorter than ——.	**d.** $\dfrac{99}{100}$ × —— is longer/shorter than ——.

4. Let s be the scaling factor. For what kind of values of s will s × \$500 be more than \$500? For what kind of values will it be less?

5. Write < , > , or = in the boxes. Fill in a number on the empty lines.

A quantity (or a number) is scaled by scaling factor s.

When s ☐ ___ , the resulting quantity is more than the original.

When s ☐ ___ , the resulting quantity is less than the original.

When s ☐ ___ , the resulting quantity is equal to the original.

6. Scaling is also the concept we use when calculating prices. Find the total cost. Use either fractions or decimals, depending on what makes most sense.

   a. Nuts cost \$8.50 per pound. You buy 1 ½ pounds.

   b. Rent is \$350 per month (30 days). You stay for 12 days.

A neat connection.

You have learned to use multiplication with equivalent fractions. →

We can write the same process this way: $\dfrac{3}{4} = \dfrac{5 \times 3}{5 \times 4} = \dfrac{15}{20}$

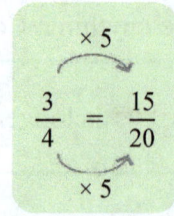

$$\overset{\times\,5}{\underset{\times\,5}{\curvearrowright}} \quad \dfrac{3}{4} = \dfrac{15}{20}$$

Notice: $\dfrac{5 \times 3}{5 \times 4}$ is the same as $\dfrac{5}{5} \times \dfrac{3}{4}$, isn't it? And $\dfrac{5}{5}$ is equal to 1.

Therefore, $\dfrac{5 \times 3}{5 \times 4}$ is actually the same as multiplying $\dfrac{3}{4}$ by 1.

So, we can make equivalent fractions by multiplying a given fraction by 1; we just write 1 in the form of a fraction (such as 3/3 or 11/11).

7. Make equivalent fractions by multiplying the given fraction by different forms of the number 1.

a. Multiply by $\dfrac{4}{4}$.	**b.** Multiply by $\dfrac{3}{3}$.	**d.** Multiply by $\dfrac{7}{7}$.
$\underline{\quad\quad} \times \dfrac{2}{3} =$	$\underline{\quad\quad} \times \dfrac{5}{9} =$	$\underline{\quad\quad} \times \dfrac{11}{12} =$

8. Kathy multiplied $\dfrac{2}{7}$ by $\dfrac{10}{10}$, and said it didn't really change anything. Heather chimed in and said, "No, it's 10 times bigger now."

a. Is Heather correct? Explain why or why not.

b. Which number is 10 times bigger than 2/7?

9. Is the result of multiplication more, less, or equal to the original number? You do not have to calculate anything. Compare, writing $<$, $>$, or $=$ in the box.

a. $\dfrac{9}{10} \times 16 \;\boxed{}\; 16$	**b.** $5\dfrac{7}{9} \times 31 \;\boxed{}\; 31$	**c.** $\dfrac{6}{6} \times 5 \;\boxed{}\; 5$
d. $\dfrac{20}{20} \times 88 \;\boxed{}\; 88$	**e.** $\dfrac{4}{5} \times \dfrac{2}{3} \;\boxed{}\; \dfrac{2}{3}$	**f.** $\dfrac{11}{4} \times 164 \;\boxed{}\; 164$
g. $\dfrac{7}{5} \times \dfrac{4}{4} \;\boxed{}\; \dfrac{7}{5}$	**h.** $2.61 \times 7 \;\boxed{}\; 7$	**i.** $0.918 \times 431 \;\boxed{}\; 431$

Fractions Are Divisions

1. You want to share *two* pies evenly between *three* people.
   How will that work? What part (fraction) of one pie will each person get?
   Use drawings to explore the situation.

2. Continue exploring these types of uneven divisions. Fill in, the best you can. If you feel
   confused, don't worry! We will look at this concept in more detail on the next page.

a. Divide 3 pies equally among four
people. In other words, solve 3 ÷ 4.

Each will get ⬜/⬜ of a pie.

b. Divide 3 protein bars equally among five
people. In other words, solve 3 ÷ 5.

Each will get ⬜/⬜ of a bar.

c. Divide 5 pies equally among six people.
In other words, solve 5 ÷ 6.

Each person will get ⬜/⬜ of a pie.

d. Divide 6 protein bars equally among
four people. In other words, solve 6 ÷ 4.

Each will get _____ bars.

e. Divide 11 pies equally among eight people. In other words, solve 11 ÷ 8.

Each person will get _____ pies.

Each fraction is also a division problem.

Example 1. If five people share two pies evenly, how much does each one get? In other words, what is **2 ÷ 5**?

The amazing answer is that $2 \div 5$ is $\frac{2}{5}$!

How can we be sure of that? Look at the picture. Can you see how 2 pies are divided into five equal parts? Each person gets two fifths of a pie.

You could also do the division this way: give each person 1 slice of *each* pie. Again, each person will get two slices, or 2/5 of a pie.

The same idea works with improper fractions: An improper fraction, too, is a division problem.

Example 2. If four people share 11 apples equally, the division $11 \div 4$ tells us how many apples each person gets. (Note: We do *not* write $4 \div 11$ — that would mean 4 apples being shared among 11 people.)

Now, $11 \div 4 = \frac{11}{4}$, which equals 2 ¾ as a mixed number. So each person gets 2 ¾ apples.

Notice that this answer, 2 ¾, is between 2 and 3. It is more than 2, and less than 3.

Note especially: $9 = 9 \div 1 = \frac{9}{1}$. Each whole number n can be written as the fraction $\frac{n}{1}$.

3. Fill in.

a. The answer to the division $3 \div 5$ is $\frac{}{}$.

b. $8 \div 21 = \frac{}{}$

c. $21 \div 100 = \frac{}{}$

d. Five people share 6 pies equally. Equation: _____ ÷ _____ = $\frac{}{}$ = $\frac{}{}$

Each person will get _____ pies.
Between which two whole numbers is the answer to this? Between _____ and _____.

e. $31 \div 8 = \frac{}{} = \frac{}{}$

The answer is between the whole numbers _____ and _____.

f. $46 \div 5 = \frac{}{} = \frac{}{}$

The answer is between the whole numbers _____ and _____.

g. If six people share 17 pizzas evenly, each person gets _____ pizzas.

The answer is between the whole numbers _____ and _____.

h. The answer to $61 \div 8$ is between the whole numbers _____ and _____.

But wait! What about remainders? Isn't the answer to 11 ÷ 4 also 2 R3?

Yes, it is. Depending on <u>context</u>, you might give the answer to 11 ÷ 4 as 2 R3 or 2 ¾ or 2.75. The way you give the answer to a division problem depends on the *type* of problem:

- Some things *cannot* be divided into parts, so you need to <u>give the answer with the remainder</u>.
- Some things *can* be divided into parts, so you <u>give the answer as a mixed number or a decimal</u>.

Example 3. A teacher shares 65 pencils equally with 26 students. How many will each get?

It does not make sense to break a pencil into parts, so we will give the answer <u>with a remainder</u>: 65 ÷ 26 = 2 R13. Each student gets two pencils, and 13 are left over.

Example 4. Twenty-six students share 65 apples equally. How many apples will each get?

$65 \div 26 = \dfrac{65}{26} = 2\dfrac{13}{26} = 2\dfrac{1}{2}$. Each student gets 2 ½ apples.

4. Rewrite each division with a remainder as a division where the answer is given as a mixed number.

a. $25 \div 8 = 3$ R1 $\dfrac{25}{8} = 3\dfrac{1}{8}$	**b.** $44 \div 5 = 8$ R4 $\dfrac{}{} = \square \dfrac{}{}$	**c.** $23 \div 2 = $ _____ R _____ $\dfrac{}{} = \square \dfrac{}{}$
d. $28 \div 3 = $ _____ R _____ $\dfrac{}{} = \square \dfrac{}{}$	**e.** $65 \div 10 = $ _____ R _____ $\dfrac{}{} = \square \dfrac{}{}$	**f.** $53 \div 9 = $ _____ R _____ $\dfrac{}{} = \square \dfrac{}{}$

5. Alison divided 15 lb of berries equally into four bags.
   How many pounds of berries did she put into one bag?

6. Seventy-five people are organized into groups of 4, as evenly as possible.
   How many and what kind of groups do they get?

7. You divide five chocolate bars equally among you and your two sisters.
   How much does each person get?

8. **a.** Between which two whole numbers is the answer to 45 ÷ 6?

   b. Find the answer to 45 ÷ 6 both as
      a mixed number and as a decimal.

9. Calculate

a. 1/12 of 15 kg	**b.** 1/4 of 7 inches

10. If nine people share a 50-pound sack of rice equally by
    weight, how many pounds of rice should each person get?

    Between which two whole numbers does your answer lie?

11. One mini-bus can seat 11 people. How many
    mini-buses do you need to transport 102 people?

12. Noah needs to pour 5 liters of juice evenly into 20 glasses.

    a. How many liters is in one glass?

    b. How many milliliters is that?

Dividing Fractions: Sharing Divisions

One meaning of division is **equal sharing.** In this lesson, we will look at sharing (dividing!) pie pieces equally among a certain number of people. This means the divisor is a <u>whole number</u>.

Example 1.

$\dfrac{9}{10} \div 3 = ??$

Think of 9/10 of a pie being divided among three people. We simply divide the number of slices, 9, by 3, and get that each person gets 3 slices.

Since the slices are tenths, each person gets 3/10 of the pie.

Each division can be **checked with a multiplication:** take the quotient (answer) times the divisor. You should get the original dividend. In this case, $\dfrac{3}{10} \times 3 = \dfrac{9}{10}$, so it checks.

1. Write a division sentence. You can color pie pieces to help. Lastly, write a multiplication to check your division.

a. $\dfrac{6}{9}$ of a pie is divided between two people.

$\dfrac{6}{9} \div 2 = \underline{}$

Check: $\underline{} \times 2 = $

b. $\dfrac{3}{5}$ of a pie is divided among three people.

$\underline{} \div 3 = \underline{}$

Check: $\underline{} \times \underline{} = $

c. $\dfrac{6}{12} \div 3 = \underline{}$

Check: $\underline{} \times \underline{} = $

d. $\dfrac{15}{20} \div 5 = \underline{}$

Check: $\underline{} \times \underline{} = $

2. Write a division sentence for each problem, and solve it.

a. There is 6/9 of a pizza left over, and three people share it equally. How much does each one get?	**b.** A cake was cut into 20 pieces, and now there are 12 pieces left. Four people share those equally. What fraction of the original cake does each person get?

Next, we will divide **unit fractions**—fractions of the form 1/*n* where *n* is a whole number.

Example 2. $\dfrac{1}{2} \div 4 = ??$ Think of 1/2 of a pie being divided equally among four people. What fractional part of the *original* pie will each person get?

Remember also that **multiplication** is the **opposite operation of division.** This means that whatever the answer is to (1/2) ÷ 4, if you multiply the answer by 4, you will get 1/2.

(See the bottom of the page for the answer.)

3. Split each unit fraction equally, and solve. Use multiplication to check. Also, look for a shortcut, and fill it in.

a.

$\dfrac{1}{3} \div 2 = $ ____

$\dfrac{}{} \times 2 = \dfrac{1}{3}$

b.

$\dfrac{1}{2} \div 5 = $

$\dfrac{}{} \times$ ____ $=$

c.

$\dfrac{1}{4} \div 2 = $

$\dfrac{}{} \times$ ____ $=$

d.

$\dfrac{1}{5} \div 3 = $

$\dfrac{}{} \times$ ____ $=$

e.

$\dfrac{1}{3} \div 3 = $

$\dfrac{}{} \times$ ____ $=$

f.

$\dfrac{1}{5} \div 2 = $

$\dfrac{}{} \times$ ____ $=$

Shortcut: $\dfrac{1}{m} \div n = \dfrac{}{}$ (where 1/*m* is a unit fraction, and *n* is a whole number)

4. Three children share 1/4 lb of chocolate equally.
   How much does each one get (in pounds)?

5. Solve.

a. $\dfrac{1}{6} \div 3 =$	**b.** $\dfrac{1}{10} \div 2 =$	**c.** $\dfrac{1}{7} \div 6 =$	**d.** $\dfrac{1}{2} \div 14 =$

Each person will get 1/8 of the original pie.

42

6. A half liter of juice is poured evenly into five glasses.

   a. How much juice is in each glass, measured in <u>liters</u>?

   b. How about in <u>milliliters</u>?

7. One morning, Joshua's gas can was only 1/8 full. He poured half of the gas into his lawn mower.

   a. How full is the gas can now?

   b. If the gas can holds 3 gallons, how much gasoline is left in the can, in gallons?

8. The job of cleaning the bathrooms at summer camp was first divided equally between Jeremy and Jenny. Jenny then amassed her a team of three other girls plus herself, and divided her part evenly among her team. What part of the job did each girl in Jenny's team do?

9. Think of equal sharing, and solve.

a. $\dfrac{12}{20} \div 2 =$	**b.** $\dfrac{8}{11} \div 4 =$	**c.** $\dfrac{8}{5} \div 4 =$	**d.** $\dfrac{2}{9} \div 2 =$

10. Write a story problem to match each division, and solve.

a. $\dfrac{1}{2} \div 3 =$	**b.** $\dfrac{6}{8} \div 2 =$

Puzzle Corner Find the missing fractions and mixed numbers.

a. $\dfrac{}{} \div 3 = \dfrac{1}{8}$	**b.** $\dfrac{}{} \div 5 = \dfrac{2}{7}$	**c.** $\dfrac{}{} \div 3 = \dfrac{4}{10}$	**d.** $\dfrac{}{} \div 6 = \dfrac{5}{8}$

Dividing Fractions: Fitting the Divisor

Besides equal sharing, we also use division to answer questions of the type: "*How many times does one number go into another?*" For example, the answer to "How many 7s are there in 336?" is found by dividing $336 \div 7$. We can use this same thought with fractions.

Example 1. What is $2 \div \frac{1}{4}$? In visual terms, how many times does ◹ fit into ⊕ ⊕ ?

To solve it, we can think this way:

- 1/4 goes into one whole four times.
- So, it goes into two wholes twice as many times, or eight times.

We can check our thinking by doing the multiplication: $8 \times \dfrac{1}{4} = \dfrac{8}{4} = 2$.

So, $2 \div \frac{1}{4} = 8$.

1. Solve. Lastly, write a multiplication that checks the division.

a. How many times does ◖ go into ◉ ◉ ? $2 \div \dfrac{1}{3} =$ Check: _____ $\times \dfrac{1}{3} =$	**b.** How many times does ◹ go into ⊕ ? $1 \div \dfrac{1}{4} =$ Check: _____ $\times \dfrac{1}{4} =$
c. How many times does ◖ go into ●●●●●● ? $6 \div \dfrac{1}{3} =$ Check:	**d.** How many times does ◹ go into ●●●●● ? $5 \div \dfrac{}{} =$ Check:
e. How many times does ⅓ go into 5? $ \div =$ Check:	**f.** How many times does ½ go into 6? $ \div =$ Check:

Shortcut: $n \div \dfrac{1}{m} = \boxed{}$ (where n is a whole number, and $1/m$ is a unit fraction)

44

2. Divide. Think, "How many times does the *divisor* go into the *dividend*?"

a. $3 \div \dfrac{1}{6} =$	**b.** $4 \div \dfrac{1}{9} =$	**c.** $4 \div \dfrac{1}{8} =$
d. $5 \div \dfrac{1}{10} =$	**e.** $7 \div \dfrac{1}{4} =$	**f.** $4 \div \dfrac{1}{10} =$

3. Write a division for each word problem, and solve. Do *not* write just the answer.

a. How many ½-meter pieces can you cut from a roll of string that is 6 meters long?

b. How many ¼-cup servings can you get from 2 cups of almonds?

c. Ben grew tomatoes that weighed ¹⁄₁₀ kg each. How many of those would he need to make 5 kg?

d. An eraser is ⅛ inches thick. How many erasers can be stacked into a 4-inch tall box?

4. (*Optional*) Try these challenges. Think, "How many times does the *divisor* go into the *dividend*?"

a. $3 \div \dfrac{3}{5} =$	**b.** $2\dfrac{5}{6} \div \dfrac{1}{6} =$	**c.** $1\dfrac{7}{8} \div \dfrac{3}{8} =$
d. $\dfrac{1}{2} \div \dfrac{1}{4} =$	**e.** $\dfrac{1}{4} \div \dfrac{1}{2} =$	**f.** $4\dfrac{1}{2} \div \dfrac{3}{4} =$

This page reviews some fraction concepts you have learned in previous lessons.

5. Let's review fraction addition and subtraction for a bit!

a. $\dfrac{5}{6} + \dfrac{3}{8}$	**b.** $\dfrac{5}{6} - \dfrac{3}{8}$
c. $3\dfrac{1}{7} + 1\dfrac{1}{2}$	**d.** $10\dfrac{2}{7} - 3\dfrac{1}{5}$

6. Jackie walks 3/8 of a mile to school every day, and the same distance back.
   What distance does she walk in a five-day school week?

7. A recipe for ginger cookies calls for 2 ¼ cups of flour. If you make
   the recipe 1 ½ times, how much flour do you need?

8. A piece of land was split between two brothers so that one got 3/5 of the land and the other got 2/5.
   Then, the brother who got more, decided to use half of his land for growing crops.
   If you multiply the two fractions, 3/5, and 1/2, what does the answer to that multiplication tell you?

Dividing Fractions: Summary

Compare the two ways to think of division		
Equal sharing, example 1 $$\frac{3}{4} \div 3 = \frac{1}{4}$$ If three people share ¾ of an apple evenly, each gets ¼ of the apple.	Equal sharing, example 2 $$7 \div 3 = \frac{7}{3} = 2\frac{1}{3}$$ If three people share 7 apples evenly, each gets 2 ⅓ apples. Notice this was a whole-number division—we studied those in the lesson "Fractions are Divisions".	Fitting one number into another $$3\frac{1}{2} \div \frac{1}{2}$$ $$\downarrow \; \downarrow \; \downarrow$$ $$3\frac{1}{2} \times 2 = 7$$ From 3 ½ lb of beef, you can get 7 half-pound servings.

1. Write a division and solve. Use the pictures and/or the shortcut to help.

a. A teacher shares 10 apples among 20 children evenly. How much does each child get?

b. A teacher shares 25 apples among 20 children evenly. How much does each child get?

c. A doctor reserves ¾ hour for each patient. How many patients could he see in 3 ¾ hours?

d. Anne has 1 ½ cups of almond flour. She divides that evenly between three batches of cookies she is making. How much almond flour goes into each batch of cookies?

e. Three people share evenly 2 2/5 pies. How much does each one get?

f. A recipe calls for ½ cup of butter, among other ingredients. How many batches of the recipe can be made if there are 3 cups of butter?

How about if there are 2 ½ cups of butter?

2. Solve.

a. $10 \div \dfrac{1}{5} =$	**b.** $4 \div 5 =$	**c.** $\dfrac{8}{9} \div 4 =$
d. $\dfrac{1}{10} \div 2 =$	**e.** $15 \div 4 =$	**f.** $6 \div \dfrac{1}{8} =$
g. $\dfrac{12}{5} \div 6 =$	**h.** $5 \div \dfrac{1}{10} =$	**i.** $\dfrac{1}{20} \div 3 =$

3. When Natalie goes jogging, she jogs for ¼ mile, then walks for ¼ mile, then again jogs for ¼ mile, and so on. How many ¼ mile stretches are there for her in a jogging track that is 2 ½ miles long?

4. Describe a situation to match each division, and solve.

a. $2 \div \dfrac{1}{2} =$
b. $\dfrac{1}{3} \div 2 =$

5. Jill makes bead necklaces that must be exactly 24 inches long. She has size SS beads, which are ⅛-inch thick, and size S beads, which are ¼-inch thick.

a. How many beads would be in a necklace made solely of SS beads?

Bead	Width
SS	⅛ in
S	¼ in

b. How many beads would be in a necklace made solely of S beads?

c. (Challenge) She also makes a necklace with the pattern SS-S-SS-S. How many of each kind of bead does she need?

Dividing Fractions: The Shortcut

(This lesson is optional.)

It may sound kind of strange, but the quickest way to calculate the answer to most fraction divisions is to **change the division into a *multiplication*.** We will explore that thought in this lesson.

How can a division be changed into a multiplication?

Think about this: We have studied how finding half of a number can be calculated by multiplying the number by 1/2. But, finding half of a number is the SAME as *dividing* the number by 2.

Similarly, to multiply a number by 1/3 will give us one third of that number. But another way to find 1/3 of any number is to divide that number by 3.

And lastly, keep in mind that the two factors in multiplication can be written in either order.

Multiplication	Division
$\frac{1}{2} \times 68 = 34$ *or* $68 \times \frac{1}{2} = 34$	$68 \div 2 = 34$
$\frac{1}{3} \times 240 = 80$ *or* $240 \times \frac{1}{3} = 80$	$240 \div 3 = 80$
$\frac{1}{6} \times 42 = 7$ *or* $42 \times \frac{1}{6} = 7$	$42 \div 6 = 7$

In a nutshell, we change a division by 3 into a multiplication by $\frac{1}{3}$. Similarly, we can change a division by 5 into a multiplication by $\frac{1}{5}$, or a division by 10 into a multiplication by $\frac{1}{10}$, ... and so on, for all unit fractions (fractions with a numerator of 1).

1. Change each division into a multiplication, and solve.

a. $30 \div 5$

↓ ↓ ↓

$30 \times \frac{1}{5} = 6$

b. $\frac{1}{9} \div 3$

↓ ↓ ↓

$\frac{1}{9} \times \frac{1}{3} = \frac{}{}$

c. $\frac{1}{4} \div 2$

↓ ↓ ↓

$\frac{}{} \times \frac{}{} = \frac{}{}$

d. $\frac{1}{7} \div 3$

↓ ↓ ↓

$\frac{}{} \times \frac{}{} = \frac{}{}$

e. $32 \div 8$

↓ ↓ ↓

f. $\frac{1}{5} \div 4$

↓ ↓ ↓

The numbers 3 and $\frac{1}{3}$ are **reciprocal numbers.** So are 5 and $\frac{1}{5}$.

Two numbers are reciprocal numbers if, when you multiply them, you get 1.

For example, the numbers $\frac{1}{9}$ and 9 are reciprocal numbers, because $\frac{1}{9} \times 9 = \frac{9}{9} = 1$.

Similarly, $\frac{2}{5}$ and $\frac{5}{2}$ are reciprocal numbers, because $\frac{2}{5} \times \frac{5}{2} = \frac{10}{10} = 1$.

Example 1. You can find a reciprocal of any fraction by "*flipping*" it (switching the numerator and the denominator). The reciprocal of $\frac{6}{13}$ is $\frac{13}{6}$.	**Example 2.** To find the reciprocal of a whole number, first write it as a fraction with a denominator of one: $27 = \frac{27}{1}$. Then, "flip" it, and you get $\frac{1}{27}$ as its reciprocal.

2. Find the reciprocals of these numbers.

a. $\frac{1}{6}$	b. $\frac{1}{100}$	c. 5	d. 21	e. $\frac{7}{8}$
f. $\frac{11}{3}$	g. $\frac{1}{42}$	h. 1	i. 13	j. $\frac{5}{6}$

The shortcut for fraction division is: **Change the division into a multiplication, and the divisor into its reciprocal.**	**Example 3.** $\frac{6}{7} \div 2$ $\downarrow \quad \downarrow \quad \downarrow$ $\frac{6}{7} \times \frac{1}{2} = \frac{6}{14} = \frac{3}{7}$

3. Change each division into a multiplication, and solve.

a. $\frac{1}{2} \div 6$ $\downarrow \quad \downarrow \quad \downarrow$	b. $\frac{1}{8} \div 5$ $\downarrow \quad \downarrow \quad \downarrow$	c. $\frac{1}{30} \div 3$ $\downarrow \quad \downarrow \quad \downarrow$
d. $\frac{4}{5} \div 2$ $\downarrow \quad \downarrow \quad \downarrow$	e. $\frac{3}{8} \div 3$ $\downarrow \quad \downarrow \quad \downarrow$	f. $\frac{8}{15} \div 4$ $\downarrow \quad \downarrow \quad \downarrow$

The shortcut also works when the divisor is a fraction (in fact, it always works... no matter what kind of numbers you have for the dividend and the divisor). Study the examples.

Example 4. $7 \div \dfrac{1}{2}$ $\downarrow \quad \downarrow \quad \downarrow$ $7 \times 2 = 14$	**Example 5.** $\dfrac{3}{4} \div \dfrac{2}{5}$ $\downarrow \quad \downarrow \quad \downarrow$ $\dfrac{3}{4} \times \dfrac{5}{2} = \dfrac{15}{8} = 1\dfrac{7}{8}$

4. Change each division into a multiplication, and solve.

a. $8 \div \dfrac{1}{2}$ $\downarrow \quad \downarrow \quad \downarrow$	**b.** $7 \div \dfrac{1}{3}$ $\downarrow \quad \downarrow \quad \downarrow$	**c.** $11 \div \dfrac{1}{5}$ $\downarrow \quad \downarrow \quad \downarrow$
d. $\dfrac{4}{5} \div \dfrac{1}{2}$	**e.** $\dfrac{1}{9} \div 3$	**f.** $\dfrac{4}{5} \div \dfrac{8}{9}$
g. $\dfrac{8}{5} \div \dfrac{2}{9}$	**h.** $\dfrac{5}{3} \div \dfrac{1}{2}$	**i.** $\dfrac{8}{7} \div 5$

Puzzle Corner	The shortcut for fraction division works also with mixed numbers. You just need to convert mixed numbers to fractions before dividing. See how well you do with these problems!

a. $\dfrac{29}{10} \div 1\dfrac{2}{5}$	**b.** $1\dfrac{7}{8} \div \dfrac{3}{4}$	**c.** $3\dfrac{5}{6} \div 1\dfrac{1}{2}$

51

Review

1. In (a), complete the simplification process. In the rest, reduce the fractions to lowest terms.

a. $\dfrac{9}{12} = \dfrac{\quad}{\quad}$	**b.** $\dfrac{56}{49} =$	**c.** $3\,\dfrac{15}{35} =$
	d. $2\,\dfrac{72}{84} =$	**e.** $\dfrac{12}{100} =$

2. Draw a picture to illustrate each multiplication, and solve.

a. $3 \times 1\dfrac{1}{3}$	**b.** $2 \times \dfrac{5}{6}$

3. (optional) Simplify before you multiply.

a. $\dfrac{7}{14} \times \dfrac{3}{12}$	**b.** $\dfrac{5}{24} \times \dfrac{12}{30}$

4. Multiply.

a. $7 \times \dfrac{2}{5}$	**b.** $\dfrac{2}{7} \times \dfrac{5}{6}$
c. $4\dfrac{3}{10} \times 4$	**d.** $1\dfrac{1}{6} \times 5\dfrac{2}{3}$

5. Figure out the side lengths of the colored rectangle from the picture. Then multiply the side lengths to find its area. <u>Check that the area you get by multiplying is the same as what you can see</u> <u>from the picture.</u>

a.

? (width) ? (height) 1 m 1 m

Side lengths: _____ m and _____ m

Area: _____ m × _____ m =

b.

1 km 1 km

Side lengths: _____ km and _____ km

Area: _____ km × _____ km =

6. Shade a rectangle inside the square so that its area can be found by the fraction multiplication.

a. $\dfrac{5}{6}$ m × $\dfrac{1}{2}$ m = _____ m^2

b. $\dfrac{2}{3}$ in × $\dfrac{1}{6}$ in = _____ in^2

7. Mary jogs 3/4 of a mile each day, five days a week.
Calculate how many miles she jogs in a week.

8. The morning after Sam's birthday, 12/20 of his birthday cake is left. He eats 2/3 of what is left. When you multiply those two fractions, what does your answer mean or tell you?

9. Sally made a rectangular blueberry pie, and cut it into 20 equal pieces. The next morning, 12/20 of it was left. Then, the dog got on the table and gobbled up 2/3 of what was left!

   a. How many pieces did the dog eat?

   b. How many pieces are left now?

   c. What fraction of the pie is left now?

10. Draw a picture to illustrate these calculations, and solve.

a. $1 \div 3$	**b.** $\dfrac{1}{2} \div 3$

11. Divide.

a. $2 \div \dfrac{1}{3}$	**b.** $4 \div \dfrac{1}{4}$	**c.** $\dfrac{1}{2} \div 5$
d. $\dfrac{1}{7} \div 3$	**e.** $9 \div 4$	**f.** $\dfrac{1}{8} \div 2$
g. $6 \div 9$	**h.** $\dfrac{6}{10} \div 3$	**i.** $\dfrac{3}{4} \div 3$

12. Solve.

a. A string that is 7 inches long is cut into four equal pieces. How long are the pieces?

b. Find four-fifths of the fraction 1/3.

c. Five people are sharing equally 11 lb of almonds. How many pounds will each get?

d. Chain costs $24 per meter, and you bought 3/4 of a meter. What was the cost?

13. Compare, writing < , > , or = in the box. You do not have to calculate anything.

a. $\dfrac{8}{9} \times 7 \,\square\, 7$	**b.** $2\dfrac{1}{11} \times 57 \,\square\, 57$	**c.** $\dfrac{7}{7} \times 13 \,\square\, 13$

14. One-third of a cake was decorated with chocolates,
one-fourth with sprinkles, and the rest with strawberry frosting.
What *part* was decorated with strawberry frosting?

15. A loaf of bread was cut into 30 slices. After a day,
5/6 of it was left. Then the family ate 1/5 of the
remaining bread. How many slices are left now?

16. You only have ¾ cup of walnuts in
the cupboard, so you decide to make
only ¾ of the recipe. How much of
of each ingredient do you need?

Brownies

3 cups sweetened carob chips
8 tablespoons olive oil
2 eggs
1/2 cup honey
1 teaspoon vanilla
3/4 cup whole wheat flour
3/4 teaspoon baking powder
1 cup walnuts or other nuts

17. **a.** Determine which sheet of paper has the greater area:
(1) a 6 ½ in by 8½ in sheet, or (2) a 5¾ in by 9 in sheet.

b. How much greater is the area of the larger sheet than
the area of the smaller sheet, in square inches?

Puzzle Corner Find the expressions that are equivalent to 5/8.

a. $\frac{3 \times 5}{6 \times 4}$	**b.** $\frac{9 \times 5}{8 \times 9}$	**c.** $\frac{5}{6} \times \frac{6}{5}$	**d.** $\frac{5}{7} \times \frac{7}{8}$	**e.** $\frac{2}{8} \times \frac{5}{3} \times \frac{3}{2}$	**f.** $\frac{4 \times 9 \times 5}{5 \times 9}$	**g.** $\frac{2 \times 5 \times 8}{8 \times 5 \times 8}$

Answer Key

Simplifying Fractions 1, pp. 7-9

1.

a. Every _three_ slices are joined together.	b. Every _three_ slices are joined together.	c. Every _three_ slices are joined together.
÷ 3 $\frac{3}{9} = \frac{1}{3}$ ÷ 3	÷ 3 $\frac{6}{9} = \frac{2}{3}$ ÷ 3	÷ 3 $\frac{3}{6} = \frac{1}{2}$ ÷ 3
d. Every _four_ slices were joined together.	e. Every _eight_ parts were joined together.	f. Every _four_ parts were joined together.
÷ 4 $\frac{4}{20} = \frac{1}{5}$ ÷ 4	÷ 8 $\frac{8}{24} = \frac{1}{3}$ ÷ 8	÷ 4 $\frac{12}{16} = \frac{3}{4}$ ÷ 4

2.

a. Every _three_ slices were joined together.	b. Every _five_ slices were joined together.	c. Every _two_ slices were joined together.	d. Every _two_ slices were joined together.
÷ 3 $\frac{3}{12} = \frac{1}{4}$ ÷ 3	÷ 5 $\frac{5}{5} = \frac{1}{1}$ ÷ 5	÷ 2 $\frac{6}{10} = \frac{3}{5}$ ÷ 2	÷ 2 $\frac{8}{14} = \frac{4}{7}$ ÷ 2

3.

a.	b.	c.
$\frac{6}{18} = \frac{1}{3}$	$\frac{8}{12} = \frac{2}{3}$	$\frac{9}{12} = \frac{3}{4}$

4.

a. $\div 2$ $\frac{6}{16} = \frac{3}{8}$ $\div 2$	b. $\div 5$ $\frac{15}{25} = \frac{3}{5}$ $\div 5$	c. $\div 4$ $\frac{28}{32} = \frac{7}{8}$ $\div 4$	d. $\div 6$ $\frac{12}{42} = \frac{2}{7}$ $\div 6$	e. $\div 9$ $\frac{18}{27} = \frac{2}{3}$ $\div 9$

5. a. It stays the same; the value does not change.
   b. Because the resulting fraction has a smaller numerator and denominator, so in that sense it is "reduced" or simpler.

6. Yes. Examples and explanations will vary; check the student's work. For example: If the numerator and the denominator don't have any common factors, then you cannot simplify the fraction. For example, 1/2, 3/7, and 11/15 cannot be simplified.

7.

a. $\frac{12}{20} = \frac{3}{5}$	b. $\frac{24}{32} = \frac{3}{4}$	c. $\frac{3}{15} = \frac{1}{5}$	d. $\frac{15}{18} = \frac{5}{6}$	e. $\frac{16}{20} = \frac{4}{5}$

8. a. 1 1/4   b. 5 1/9   c. 7 1/4   d. 3 2/7

9. a. cannot simplify   b. 3 1/7   c. cannot simplify   d. 2/11
   e. 1/2   f. cannot simplify   g. cannot simplify   h. 3/7

10. What part of the total time is the warm-up time? 10/60 = <u>1/6</u>
    What part of the total time is the actual running practice time? 40/60 = <u>2/3</u>

Simplifying Fractions 2, pp. 10-12

1.

a. $\div 10$ $\div 4$ $\frac{40}{120} = \frac{4}{12} = \frac{1}{3}$ $\div 10$ $\div 4$	b. $\div 5$ $\div 3$ $\frac{75}{105} = \frac{15}{21} = \frac{5}{7}$ $\div 5$ $\div 3$	c. $\div 2$ $\div 7$ $\frac{42}{98} = \frac{21}{49} = \frac{3}{7}$ $\div 2$ $\div 7$
You could simplify in one step if you divided by <u>40.</u>	You could simplify in one step if you divided by <u>15.</u>	You could simplify in one step if you divided by <u>14.</u>

2. a. 3/10   b. 3/8   c. 5/14
   d. 5 3/4   e. 3 2/7   f. 7 3/5

3. a. 54/36 = 3/2 = 1 1/2
   b. 64/48 = 4/3 = 1 1/3
   c. 56/49 = 8/7 = 1 1/7

4. a. cannot simplify   b. 2 3/4   c. cannot simplify
   d. 4 2/5   e. 1 3/5   f. 2 5/11 (cannot simplify)

5. a. 21/14 + 12/14 = 33/14 = 2 5/14   b. 15/8 − 3/10 = 63/40 = 1 23/40   c. 5 5/9 + 3 7/12 = 329/36 = 9 5/36

Simplifying Fractions 2, cont.

6.

7. Who got it right? <u>Nancy and Jerry got it right.</u> Who did not? <u>Mark did not.</u>
Why? <u>Because Mark did not reduce it to the lowest terms. His answer was 8/10, which can still be simplified to 4/5.</u>

8. a. 3/4   b. 600 pixels

9.

$\frac{3}{4}$	$\frac{2}{5}$	$\frac{1}{2}$	$\frac{2}{7}$		$\frac{1}{4}$	$\frac{3}{5}$	$\frac{1}{2}$		$\frac{1}{4}$	$\frac{2}{3}$	$\frac{1}{6}$	$\frac{1}{4}$	$\frac{2}{7}$	$\frac{1}{3}$		$\frac{1}{3}$	$\frac{3}{4}$	$\frac{5}{6}$	$\frac{3}{10}$	$\frac{3}{10}$	$\frac{1}{2}$	$\frac{3}{7}$	
T	H	E	Y		A	R	E		A	L	W	A	Y	S		S	T	U	F	F	E	D	.

Because

Multiply Fractions and Whole Numbers 1, pp. 13-14

1.

a. $\underline{3} \times \frac{4}{5} = \frac{12}{5}$	b. $3 \times \frac{7}{9} = \frac{21}{9}$	c. $2 \times \frac{7}{8} = \frac{14}{8}$

2.

a. $3 \times \frac{7}{10} = \frac{21}{10} = 2\frac{1}{10}$	b. $4 \times \frac{7}{9} = \frac{28}{9} = 3\frac{1}{9}$	c. $3 \times \frac{5}{8} = \frac{15}{8} = 1\frac{7}{8}$

3. b. 28/10 = 2 4/5   c. 22/20 = 1 2/20 = 1 1/10   d. 18/15 = 6/5 = 1 1/5

4. 4 × (3/8 L) = 12/8 L = 1 4/8 L = 1 1/2 L

5.

a. 30/6 = 5	b. 42/100 = 21/50
c. 16/12 = 4/3 = 1 1/3	d. 70/100 = 7/10
e. 90/20 = 18/4 = 4 1/2	f. 49/15 = 3 4/15

6.

> **Brownies**
>
> 2 1/4 cups of butter
> 4 1/2 cups of brown sugar
> 12 eggs
> 3 3/4 cups of cocoa powder
> 1 1/2 cups of flour
> 6 tsp vanilla

7. a. Most students used between 1/2 and 1 1/4 hours for housework and chores.
   b. Four students.
   c. Multiply the average by the number of students to get the total:
      20 × (7/8) = 140/8 = 17 4/8 = 17 1/2 hours

Multiplying Fractions and Whole Numbers 2, pp. 15-16

1.

a. $\frac{1}{2} \times 60 = 30$	b. $\frac{1}{3} \times 150 = 50$	c. $\frac{1}{8} \times 24 = 3$
d. $\frac{4}{5} \times 100 = 80$	e. $\frac{2}{3} \times 36 = 24$	f. $\frac{5}{6} \times 30 = 25$

2. You can take the answer to the top problem and multiply it by 3, 2, or 8 (respectively) to get the answer to the bottom one.

a. $\frac{1}{4} \times 60 = 15$	b. $\frac{1}{5} \times 45 = 9$	c. $\frac{1}{9} \times 180 = 20$
$\frac{3}{4} \times 60 = 45$	$\frac{2}{5} \times 45 = 18$	$\frac{8}{9} \times 180 = 160$

3. a. 15 km   b. $24
   c. 350 kg   d. 24 in

4. a. 14 lb   b. 80 km

5. Answers will vary. Check the student's answer. For example:
   You agreed to pay 2/3 of the cost of a $600 bike.
   How much is your share? It is $400.

6. 2/5 of 400 cm = 160 cm or 1.6 m.

7. a. Since 1/3 of $81 is $27, Janet got $54
   and Sandy got $27.

   b. We would need to use long division, and divide $80.00 by 3 (see the division on
   the right). The division will not end, so the answer needs to be rounded. Now,
   Sandy would get $26.67 and Janet would get the rest, or $53.33.

```
     2 6.6 6 6
  3)8 0.0 0 0
   -6
    2 0
   -1 8
     2 0
    - 1 8
      2 0
     -1 8
       2 0
```

Multiply Fractions by Fractions 1, pp. 17-19

1.

a. $\frac{1}{2}$ of ⬤ is ⬤ $\frac{1}{2} \times \frac{1}{2} = \frac{1}{4}$	b. $\frac{1}{2}$ of ⬤ is ⬤ $\frac{1}{2} \times \frac{1}{3} = \frac{1}{6}$	c. $\frac{1}{2}$ of ⬤ is ⬤ $\frac{1}{2} \times \frac{1}{4} = \frac{1}{8}$
d. $\frac{1}{3}$ of ⬤ is ⬤ $\frac{1}{3} \times \frac{1}{2} = \frac{1}{6}$	e. $\frac{1}{3}$ of ⬤ is ⬤ $\frac{1}{3} \times \frac{1}{3} = \frac{1}{9}$	f. $\frac{1}{3}$ of ⬤ is ⬤ $\frac{1}{3} \times \frac{1}{4} = \frac{1}{12}$
g. $\frac{1}{4}$ of ⬤ is ⬤ $\frac{1}{4} \times \frac{1}{2} = \frac{1}{8}$	h. $\frac{1}{4}$ of ⬤ is ⬤ $\frac{1}{4} \times \frac{1}{3} = \frac{1}{12}$	i. $\frac{1}{4}$ of ⬤ is ⬤ $\frac{1}{4} \times \frac{1}{4} = \frac{1}{16}$

2. <u>You can simply multiply the denominators.</u> For example, $\frac{1}{5} \times \frac{1}{2} = \frac{1}{10}$ and $\frac{1}{4} \times \frac{1}{6} = \frac{1}{24}$.

Multiply Fractions by Fractions 1, cont.

3. a. 1/18   b. 1/39   c. 1/100

4.

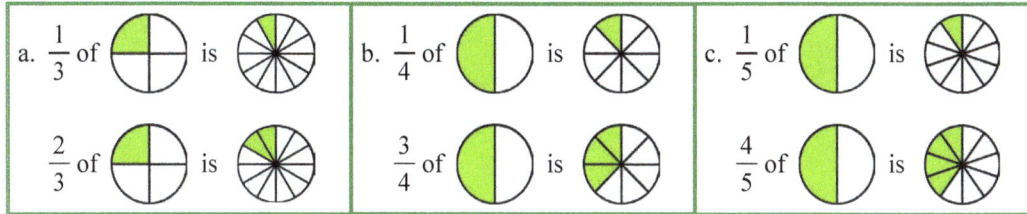

5. a. $\frac{1}{2} \times \frac{2}{3} = \frac{1}{3}$

  b. It is 1/3 of the *original* pizza.

6. a. 1/35, 2/35   b. 1/24, 5/24   c. 1/24, 3/24

7. a. 2/27   b. 11/72
   c. 1/13   d. 6
   e. 4/21   f. 7 1/7

Multiply Fractions by Fractions 2, pp. 20-21

1. a. (1/4) × (2/3) = 2/12 = 1/6
   b. 2/12 or 1/6 of the original pizza.
   c. 1/12. When 1/4 of the pizza was left, that was 3/12. Marie ate 2/12, so 1/12 of the original pizza is now left.

2. a. 3/8

   b. (1/2) × (3/8) = 3/16.
     At first, Theresa painted 5/8 of the room:

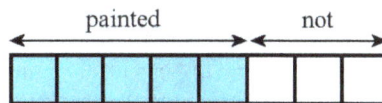

     Later, she painted 1/2 of what was left.

   Now there is still 3/16 of the room still left to paint.
   In the bar model, we see 1/8 and half of 1/8 of the room left to paint, and this is equal to 3/16.

3. See the recipe on the right.

4. a. 21/32   b. 1 3/25
   c. 9/25   d. 7 1/5
   e. 21 3/7   f. 1 7/11

5. a. 1/3

   b.
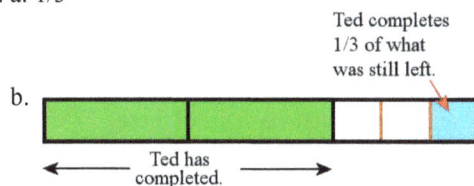

   (1/3) × (1/3) = 1/9. Ted completes 1/9 of the original job now.
   Now, 2/9 of the job is still not done, and 7/9 of it is completed.

Carob Brownies (1/3 recipe)

1 cup sweetened carob chips
2 2/3 tablespoons olive oil
1 (small) egg
1/6 cup honey
1/3 teaspoon vanilla
1/4 cup whole wheat flour
1/4 teaspoon baking powder
1/3 cup walnuts or other nuts

6. a. Thirty guests that drink 30 servings means Alison needs to make the recipe 6 times. That means
      she needs 6 × 1/4 = 6/4 = 1 1/2 cups of ground coffee.
   b. Fifty guests that each drink two servings means 100 servings, so Alison needs to make the recipe 20 times.
      That means she needs 20 × 1/4 = 20/4 = 5 cups of ground coffee.

Puzzle corner: a. 1/6   b. 5/4   c. 3/4

61

Fraction Multiplication and Area, pp. 22-27

1.

a. Side lengths: $\dfrac{1}{2}$ m and $\dfrac{3}{4}$ m

Area (from the picture): $\dfrac{3}{8}$ m^2

b. Side lengths: $\dfrac{2}{3}$ in and $\dfrac{2}{3}$ in

Area (from the picture): $\dfrac{4}{9}$ in^2

2.

a. Side lengths: $\dfrac{1}{3}$ m and $\dfrac{3}{4}$ m

Area (by multiplication):

$\dfrac{1}{3}$ m $\times$ $\dfrac{3}{4}$ m $=$ $\dfrac{3}{12}$ m^2

b. Side lengths: $\dfrac{1}{3}$ in and $\dfrac{2}{3}$ in

Area (by multiplication):

$\dfrac{1}{3}$ in $\times$ $\dfrac{2}{3}$ in $=$ $\dfrac{2}{9}$ in^2

c. Side lengths: $\dfrac{2}{3}$ m and $\dfrac{3}{4}$ m

Area (by multiplication):

$\dfrac{2}{3}$ m $\times$ $\dfrac{3}{4}$ m $=$ $\dfrac{6}{12}$ m^2

d. Side lengths: $\dfrac{3}{5}$ km and $\dfrac{3}{4}$ km

Area (by multiplication):

$\dfrac{3}{5}$ km $\times$ $\dfrac{3}{4}$ km $=$ $\dfrac{9}{20}$ km^2

3. The coloring may vary, but the students' pictures should either look like these, or be rotated versions of these.

a. $\dfrac{1}{4}$ m $\times$ $\dfrac{1}{2}$ m $=$ $\dfrac{1}{8}$ m^2

b. $\dfrac{1}{2}$ in $\times$ $\dfrac{4}{6}$ in $=$ $\dfrac{4}{12}$ in^2

c. $\dfrac{3}{4}$ ft $\times$ $\dfrac{2}{7}$ ft $=$ $\dfrac{6}{28}$ ft^2

d. $\dfrac{3}{5}$ km $\times$ $\dfrac{5}{6}$ km $=$ $\dfrac{15}{30}$ km^2

Fraction Multiplication and Area, cont.

4. In each problem, the factors may also be written in the other order.

a. 1/3 m

1/3 m

Area: $\dfrac{1}{3}$ m $\times$ $\dfrac{1}{3}$ m $=$ $\dfrac{1}{9}$ m^2

b. 1/5 m

1/3 m

Area: $\dfrac{1}{3}$ m $\times$ $\dfrac{1}{5}$ m $=$ $\dfrac{1}{15}$ m^2

c. 1/5 m

1/2 m

Area: $\dfrac{1}{5}$ m $\times$ $\dfrac{1}{2}$ m $=$ $\dfrac{1}{10}$ m^2

d. 1/4 m

1/4 m

Area: $\dfrac{1}{4}$ m $\times$ $\dfrac{1}{4}$ m $=$ $\dfrac{1}{16}$ m^2

5. In each problem, the factors may also be written in the other order.

a. 3/4 m

1/2 m

Area: $\dfrac{3}{4}$ m $\times$ $\dfrac{1}{2}$ m $=$ $\dfrac{3}{8}$ m^2

b. 2/5 m

3/4 m

Area: $\dfrac{2}{5}$ m $\times$ $\dfrac{3}{4}$ m $=$ $\dfrac{6}{20}$ m^2

c. 2/3 m

2/3 m

Area: $\dfrac{2}{3}$ m $\times$ $\dfrac{2}{3}$ m $=$ $\dfrac{4}{9}$ m^2

d. 3/5 m

1/2 m

Area: $\dfrac{3}{5}$ m $\times$ $\dfrac{1}{2}$ m $=$ $\dfrac{3}{10}$ m^2

e. 3/4 m

3/4 m

Area: $\dfrac{3}{4}$ m $\times$ $\dfrac{3}{4}$ m $=$ $\dfrac{9}{16}$ m^2

f. 5/6 m

1/2 m

Area: $\dfrac{5}{6}$ m $\times$ $\dfrac{1}{2}$ m $=$ $\dfrac{5}{12}$ m^2

6. In each problem, the factors may also be written in the other order.

a. $\dfrac{3}{5} \times \dfrac{3}{4} = \dfrac{9}{20}$　　　b. $\dfrac{3}{5} \times \dfrac{2}{3} = \dfrac{6}{15}$

c. $\dfrac{2}{3} \times \dfrac{2}{5} = \dfrac{4}{15}$　　　d. $\dfrac{2}{2} \times \dfrac{2}{4} = \dfrac{4}{8}$

7. a. See the image on the right (not to scale). Its area is 1 square inch.
   b. Check the students' work.
   c. The area is (3/4 in) × (5/8 in) = 15/32 square inches. It is (slightly) less than half square inch.

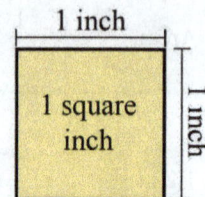

8. The area is 3 km × (1/2 km) = 1 ½ km^2.

9. a. Using fractions: (3/10 cm) × (7/10 cm) = 21/100 cm^2.
   b. Using decimals: 0.3 cm × 0.7 cm = 0.21 cm^2.

10. a. The area of one stamp is (7/8 in) × (3/4 in) = 21/32 in^2.
    The area of six stamps is 6 × (21/32 in^2) = 126/32 in^2 = 63/16 in^2 = 3 15/16 in^2.

    b. The area of the envelope is 40 square inches. The stamps cover about 1/10 of it (the area of the stamps is very close to 4 square inches).

Puzzle corner: The area of the square is (7/8 mile) × (7/8 mile) = 49/64 miles^2.
The area of the rectangle is 1/4 mile × 3 miles = 3/4 mi^2. To compare the two, we will write the latter area with a denominator of 64: 3/4 mi^2 = 48/64 mi^2 . The 7/8-mile by 7/8-mile square has a larger area than the rectangle, by 1/64 square mile.

Simplifying Before Multiplying, pp. 28-30

1.

	a.		b.		c.		d.
	$\dfrac{\overset{7}{\cancel{14}}}{\underset{8}{\cancel{16}}} = \dfrac{7}{8}$		$\dfrac{\overset{11}{\cancel{33}}}{\underset{9}{\cancel{27}}} = \dfrac{11}{9}$		$\dfrac{\overset{6}{\cancel{12}}}{\underset{13}{\cancel{26}}} = \dfrac{6}{13}$		$\dfrac{\overset{3}{\cancel{9}}}{\underset{11}{\cancel{33}}} = \dfrac{3}{11}$

2. Students don't need to write the intermediate multiplication step. For clarity, it is shown here in its entirety.

a. $\dfrac{\overset{3}{\cancel{6}}}{\underset{5}{\cancel{10}}} \times \dfrac{\overset{1}{\cancel{2}}}{\underset{7}{\cancel{14}}} = \dfrac{3 \times 1}{5 \times 7} = \dfrac{3}{35}$	b. $\dfrac{\overset{1}{\cancel{2}}}{4} \times \dfrac{\overset{1}{\cancel{3}}}{\underset{5}{\cancel{15}}} = \dfrac{1 \times 1}{2 \times 5} = \dfrac{1}{10}$
c. $\dfrac{\overset{2}{\cancel{8}}}{\underset{3}{\cancel{12}}} \times \dfrac{1}{2} = \dfrac{2 \times 1}{3 \times 2} = \dfrac{2}{6} = \dfrac{1}{3}$	d. $\dfrac{\overset{1}{\cancel{8}}}{\underset{4}{\cancel{32}}} \times \dfrac{\overset{2}{\cancel{14}}}{\underset{3}{\cancel{21}}} = \dfrac{1 \times 2}{4 \times 3} = \dfrac{2}{12} = \dfrac{1}{6}$
e. $\dfrac{\overset{2}{\cancel{6}}}{\underset{5}{\cancel{15}}} \times \dfrac{\overset{2}{\cancel{6}}}{\underset{3}{\cancel{9}}} = \dfrac{2 \times 2}{5 \times 3} = \dfrac{4}{15}$	f. $\dfrac{\overset{3}{\cancel{27}}}{\underset{5}{\cancel{45}}} \times \dfrac{\overset{3}{\cancel{21}}}{\underset{7}{\cancel{49}}} = \dfrac{3 \times 3}{5 \times 7} = \dfrac{9}{35}$

Simplifying Before Multiplying, cont.

3. Students don't need to write the intermediate multiplication step. For clarity, it is shown here in its entirety.

a. $\dfrac{8}{\overset{\ }{\underset{3}{\cancel{9}}}} \times \dfrac{\overset{2}{\cancel{6}}}{11} = \dfrac{8 \times 2}{3 \times 11} = \dfrac{16}{33}$	b. $\dfrac{3}{\underset{2}{\cancel{16}}} \times \dfrac{\overset{1}{\cancel{8}}}{5} = \dfrac{3 \times 1}{2 \times 5} = \dfrac{3}{10}$	c. $\dfrac{\overset{1}{\cancel{4}}}{7} \times \dfrac{1}{\underset{3}{\cancel{12}}} = \dfrac{1 \times 1}{7 \times 3} = \dfrac{1}{21}$

4. Students don't need to write the intermediate multiplication step. For clarity, it is shown here in its entirety.

a. $\dfrac{\overset{1}{\cancel{7}}}{\underset{4}{\cancel{8}}} \times \dfrac{\overset{1}{\cancel{2}}}{\underset{1}{\cancel{7}}} = \dfrac{1 \times 1}{4 \times 1} = \dfrac{1}{4}$	b. $\dfrac{\overset{1}{\cancel{3}}}{\underset{1}{\cancel{5}}} \times \dfrac{\overset{1}{\cancel{5}}}{\underset{2}{\cancel{6}}} = \dfrac{1 \times 1}{1 \times 2} = \dfrac{1}{2}$	c. $\dfrac{\overset{1}{\cancel{5}}}{\underset{3}{\cancel{12}}} \times \dfrac{\overset{1}{\cancel{4}}}{\underset{2}{\cancel{10}}} = \dfrac{1 \times 1}{3 \times 2} = \dfrac{1}{6}$
d. $\dfrac{\overset{1}{\cancel{9}}}{\underset{5}{\cancel{15}}} \times \dfrac{\overset{1}{\cancel{3}}}{\underset{2}{\cancel{18}}} = \dfrac{1 \times 1}{5 \times 2} = \dfrac{1}{10}$	e. $\dfrac{\overset{2}{\cancel{8}}}{11} \times \dfrac{3}{\underset{1}{\cancel{4}}} = \dfrac{2 \times 3}{11 \times 1} = \dfrac{6}{11}$	f. $\dfrac{\overset{4}{\cancel{12}}}{\underset{25}{\cancel{100}}} \times \dfrac{\overset{1}{\cancel{4}}}{\underset{5}{\cancel{15}}} = \dfrac{4 \times 1}{25 \times 5} = \dfrac{4}{125}$

5. It is 25. The student response for the reason will vary. For example: because the 36s cancel each other out. Or, because it is both a division and a multiplication by 36.

6.

a. $\dfrac{82}{\underset{1}{\cancel{77}}} \times \dfrac{\overset{1}{\cancel{77}}}{1} = 82$	b. $\dfrac{\overset{1}{\cancel{13}}}{1} \times \dfrac{49}{\underset{1}{\cancel{13}}} = 49$	c. $\dfrac{5}{\underset{1}{\cancel{6}}} \times \dfrac{\overset{4}{\cancel{24}}}{1} = 20$	d. $\dfrac{\overset{6}{\cancel{54}}}{1} \times \dfrac{2}{\underset{1}{\cancel{9}}} = 12$

7. A stack of eight is $8 \times (3/8 \text{ in}) = 3$ in tall. A stack of twenty is $20 \times (3/8 \text{ in}) = 5 \times (3/2 \text{ in}) = 15/2$ in $= 7 \ 1/2$ in tall.

8. $52 \times (3/4 \text{ kg}) = 13 \times 4 \times (3/4 \text{ kg}) = 13 \times 3$ kg $= 39$ kg.

9. It does not change the final answer at all.

Puzzle corner. The simplifications can be done in various manners, even where some are done before, and some after the multiplication. The final answer should not vary.

a. $\dfrac{\overset{1}{\cancel{3}}}{\underset{1}{\cancel{5}}} \times \dfrac{1}{7} \times \dfrac{\overset{1}{\cancel{5}}}{\underset{2}{\cancel{6}}} = \dfrac{1}{14}$	b. $\dfrac{\overset{1}{\cancel{7}}}{\underset{2}{\cancel{12}}} \times \dfrac{3}{5} \times \dfrac{\overset{1}{\cancel{6}}}{\underset{1}{\cancel{7}}} = \dfrac{3}{10}$	c. $\dfrac{1}{\underset{2 \ 1}{\cancel{12}}} \times \dfrac{\overset{2}{\cancel{4}}}{3} \times \dfrac{\overset{1}{\cancel{6}}}{7} = \dfrac{2}{21}$
d. $\dfrac{9}{\underset{2}{\cancel{10}}} \times \dfrac{\overset{1}{\cancel{5}}}{\underset{1}{\cancel{2}}} \times \dfrac{\overset{1}{\cancel{2}}}{7} = \dfrac{9}{14}$	e. $\dfrac{\overset{1}{\cancel{4}}}{\underset{1}{\cancel{5}}} \times \dfrac{\overset{3}{\cancel{9}}}{8} \times \dfrac{\overset{2 \ 1}{\cancel{10}}}{\underset{6 \ 3 \ 1}{\cancel{24}}} = \dfrac{3}{8}$	f. $\dfrac{\overset{1}{\cancel{2}}}{\underset{3}{\cancel{9}}} \times \dfrac{\overset{3 \ 1}{\cancel{6}}}{\underset{1}{\cancel{7}}} \times \dfrac{\overset{1}{\cancel{7}}}{\underset{4 \ 2}{\cancel{8}}} = \dfrac{1}{6}$

1. Note: it is possible to simplify before or after multiplying, or only after converting the answer into a mixed number. Any one of those will yield the correct answer.

a. $2\frac{1}{4} \times 1\frac{1}{2}$ $\downarrow \qquad \downarrow$ $\frac{9}{4} \times \frac{3}{2} = \frac{27}{8} = 3\frac{3}{8}$	b. $5\frac{1}{5} \times \frac{1}{6}$ $\downarrow \qquad \downarrow$ $\frac{26}{5} \times \frac{1}{6} = \frac{26}{30} = \frac{13}{15}$
c. $4\frac{1}{2} \times 1\frac{1}{5}$ $\downarrow \qquad \downarrow$ $\frac{9}{2} \times \frac{6}{5} = \frac{54}{10} = \frac{27}{5} = 5\frac{2}{5}$	d. $3\frac{1}{3} \times 2\frac{1}{10}$ $\downarrow \qquad \downarrow$ $\frac{10}{3} \times \frac{21}{10} = \frac{\overset{1}{\cancel{10}}}{\underset{1}{\cancel{3}}} \times \frac{\overset{7}{\cancel{21}}}{\underset{1}{\cancel{10}}} = 7$

2. a. We need to find its area. The area is $(5\frac{1}{2}$ ft$) \times (7\frac{1}{2}$ ft$) = (11/2$ ft$) \times (15/2$ ft$) = 165/4$ ft$^2 = 41\frac{1}{4}$ ft^2.

 b. The area of the room is 12 ft × 20 ft = 240 ft^2. The carpet is about 40 ft^2, so it covers about 1/6 of the floor.

3. The anonymous student's method omits areas 2 and 3. It only accounts for area 1 (2 × 1) and for area 4 ($\frac{1}{2} \times \frac{1}{2}$). In reality, the calculation $(2\frac{1}{2}) \times (2\frac{1}{2})$ can be broken down into four parts, like the illustration shows, and the total is 3 ¾ square units.

4.

Cheeseball	
3	2 packages cream cheese
3 3/4	2 ½ cups shredded Cheddar cheese
2 1/4	1 ½ cups chopped pecans
1 1/2	1 teaspoon grated onion

5. a. 14 2/3   b. 19/27
 c. 18   d. 1 37/40
 e. 7 7/10   e. 3 3/20

6. a. The area is 8 ½ in × 11 in = 17/2 in × 11 in = 187/2 in^2 = 93 ½ in^2.

 b. The real writing area then becomes 7 ½ in × 10 in = 15/2 in × 10 in = 75 in^2.

1.

a.	b.	c.
$\frac{1}{2} \times$ ——— = —— $\frac{1}{2} \times$ 50 px = _25_ px	$\frac{1}{4} \times$ ——— = _ $\frac{1}{4} \times$ 40 px = _10_ px	$\frac{5}{8} \times$ 400 km = _250 km_ $2\frac{5}{8} \times$ 400 km = _1,050 km_
$1\frac{1}{2} \times$ ——— = ——— $1\frac{1}{2} \times$ 50 px = _75_ px	$2\frac{1}{4} \times$ ——— = ——— $2\frac{1}{4} \times$ 40 px = _90_ px	**d.** $\frac{3}{5} \times$ \$600 = _\$360_ $3\frac{3}{5} \times$ \$600 = _\$2,160_

2. a. A number that is slightly less than 1, such as 0.9 or 0.96 or 0.89. (It could be a fraction, too.)
   b. 3300 by 2200 pixels

3. a. longer   b. shorter
   c. longer   d. shorter

4. When s is greater than 1, $s \times$ \$500 will be more than \$500.
   When s is less than 1, $s \times$ \$500 will be less than \$500.

5.

> A quantity (or a number) is scaled by scaling factor s.
>
> When $s > 1$, the resulting quantity is more than the original.
>
> When $s < 1$, the resulting quantity is less than the original.
>
> When $s = 1$, the resulting quantity is equal to the original.

6. a. 1.5 × 8.5 = 12.75. The nuts will cost \$12.75. You could use fraction multiplication, too, but it is more cumbersome:
   $1\frac{1}{2} \times 8\frac{1}{2} = (3/2) \times (17/2) = 51/4 = 12\ 3/4$.

   b. You stay 12/30 of a month, which is 2/5 of a month. The rent is (2/5) × \$350 = \$140.

7.

a.	b.	c.
$\frac{4}{4} \times \frac{2}{3} = \frac{8}{12}$	$\frac{3}{3} \times \frac{5}{9} = \frac{15}{27}$	$\frac{7}{7} \times \frac{11}{12} = \frac{77}{84}$

8. a. No, Heather is not correct. Since 10/10 = 1, Kathy simply multiplied 2/7 by one, and it did not change its value.
   b. 10 × (2/7) = 20/7 = 2 6/7

9. a. <   b. >   c. =
   d. =   e. <   f. >
   g. =   h. >   i. <

1. Answers will vary. Check the student's answer and drawings. For example:

   Each person will get 2/3 of the pie.

2.

a. Each will get $\frac{3}{4}$ of a pie.	b. Each will get $\frac{3}{5}$ of a bar.

c. Each will get $\frac{5}{6}$ of a pie.

d. Each will get 1 ½ bars (which, 6/4 = 1 2/4 = 1 ½).

e.

   Each person will get $1\frac{3}{8}$ pies.

3.

a. The answer to the division 3 ÷ 5 is $\frac{3}{5}$.	b. $8 \div 21 = \frac{8}{21}$	c. $21 \div 100 = \frac{21}{100}$

d. Five people share 6 pies equally. Equation: $\underline{6} \div \underline{5} = \frac{6}{5} = 1\frac{1}{5}$

Each person will get $\underline{1\ 1/5}$ pies.

Between what two whole numbers is the answer to this? Between $\underline{1}$ and $\underline{2}$.

e. $31 \div 8 = \frac{31}{8} = 3\frac{7}{8}$ The answer is between the whole numbers $\underline{3}$ and $\underline{4}$.	f. $46 \div 5 = \frac{46}{5} = 9\frac{1}{5}$ The answer is between the whole numbers $\underline{9}$ and $\underline{10}$.

g. If six people share 17 pizzas evenly, each person gets $\underline{2\ 5/6}$ pizzas. The answer is between the whole numbers $\underline{2}$ and $\underline{3}$.

h. The answer to 61 ÷ 8 is between the whole numbers $\underline{7}$ and $\underline{8}$.

4.

a. $25 \div 8 = 3$ R1 $\frac{25}{8} = 3\frac{1}{8}$	b. $44 \div 5 = 8$ R4 $\frac{44}{5} = 8\frac{4}{5}$	c. $23 \div 2 = 11$ R1 $\frac{23}{2} = 11\frac{1}{2}$
d. $28 \div 3 = 9$ R1 $\frac{28}{3} = 9\frac{1}{3}$	e. $65 \div 10 = 6$ R5 $\frac{65}{10} = 6\frac{5}{10}$	f. $53 \div 9 = 5$ R8 $\frac{53}{9} = 5\frac{8}{9}$

Fractions Are Divisions, cont.

5. She puts 15 lb ÷ 4 = 15/4 lb = <u>3 3/4 lb</u> into one bag.

6. Since 75 ÷ 4 = 18 R3, they will get 18 groups of 4, and one group of 3.

7. Each person gets 5/3 = 1 2/3 chocolate bars.

8. a. Between 7 and 8.

   b. The answer as a mixed number is 45/6 = 7 3/6 = <u>7 1/2</u>. As a decimal, it is 7.5.

9. a. 15 kg ÷ 12 = 15/12 kg = 1 3/12 kg = 1 1/4 kg
   b. 7 in ÷ 4 = 7/4 in = 1 3/4 in.

10. 50 lb ÷ 9 = 50/9 lb = 5 5/9 lb. The answer is between 5 and 6 pounds.

11. 102 ÷ 11 = 9 R3. You will need 10 minibuses.

12. a. 5 L ÷ 20 = 5/20 L = 1/4 L.
    b. 1/4 L is 250 ml.

Dividing Fractions: Sharing Divisions, pp. 41-43

1. (The colors may not show correctly if this is printed in black and white.)

a. $\dfrac{6}{9} \div 2 = \dfrac{3}{9}$

Check: $\dfrac{3}{9} \times 2 = \dfrac{6}{9}$

b. $\dfrac{3}{5} \div 3 = \dfrac{1}{5}$

Check: $\dfrac{1}{5} \times 3 = \dfrac{3}{5}$

c. $\dfrac{6}{12} \div 3 = \dfrac{2}{12}$

Check: $\dfrac{2}{12} \times 3 = \dfrac{6}{12}$

d. $\dfrac{15}{20} \div 5 = \dfrac{3}{20}$

Check: $\dfrac{3}{10} \times 5 = \dfrac{15}{20}$

2. a. (6/9) ÷ 3 = 2/9. Each person gets 2/9 of the pizza.
   b. (12/20) ÷ 4 = 3/20. Each person gets 3/20 of the original cake.

Teaching box. (1/2) ÷ 4 = <u>1/8</u>. If you divide even the white half into four parts, like this: , it can be easier for students to see that we are talking about eighths.

3. If you divide the "empty" or white parts also into equal parts, it is easier to see what fraction is meant.

a. $\dfrac{1}{3} \div 2 = \dfrac{1}{6}$ $\dfrac{1}{6} \times 2 = \dfrac{2}{6} = \dfrac{1}{3}$	b. $\dfrac{1}{2} \div 5 = \dfrac{1}{10}$ $\dfrac{1}{10} \times 5 = \dfrac{5}{10} = \dfrac{1}{2}$	c. $\dfrac{1}{4} \div 2 = \dfrac{1}{8}$ $\dfrac{1}{8} \times 2 = \dfrac{2}{8} = \dfrac{1}{4}$
d. $\dfrac{1}{5} \div 3 = \dfrac{1}{15}$ $\dfrac{1}{15} \times 3 = \dfrac{3}{15} = \dfrac{1}{5}$	e. $\dfrac{1}{3} \div 3 = \dfrac{1}{9}$ $\dfrac{1}{9} \times 3 = \dfrac{3}{9} = \dfrac{1}{3}$	f. $\dfrac{1}{5} \div 2 = \dfrac{1}{10}$ $\dfrac{1}{10} \times 2 = \dfrac{2}{10} = \dfrac{1}{5}$

Shortcut: $\dfrac{1}{m} \div n = \dfrac{1}{mn}$ (where $1/m$ is a unit fraction, and n is a whole number)

4. Each child gets 1/12 lb, because (1/4 lb) ÷ 3 = 1/12 lb.

5. a. 1/18   b. 1/20   c. 1/42   d. 1/28

6. a. (1/2 L) ÷ 5 = 1/10 L.
   b. One liter is 1,000 ml, so 1/10 liter is 100 ml. There is <u>100 ml</u> of juice in each glass (not that much!).

7. a. 1/16 full
   b. 3 gallons divided by 16 = <u>3/16 gallon</u>

8. 1/8 of the job. Jenny's part was to do 1/2 of the job, and that got divided among 4 girls. (1/2) ÷ 4 = 1/8.

9. a. 6/20 = 3/10   b. 2/11   c. 2/5   d. 1/9

10. The student's story problems will vary. The solutions should be the same. For example:

   a. There is half of a pizza left, and three people share it equally. How much does each one get? Solution: (1/2) ÷ 3 = 1/6. Each person gets one-sixth of a pizza.

   b. The pitcher is 6/8 full of juice. Mary and Mia share it equally. How much does each girl get? Solution: (6/8) ÷ 2 = 3/8. Each girl gets 3/8 of the pitcher of juice.

Puzzle corner. a. 3/8   b. 1 3/7 (because 5 × (2/7) = 10/7)   c. 1 2/10   d. 2 3/4 (because 6 × (5/8) = 30/8)

Dividing Fractions: Fitting the Divisor, pp. 44-46

1.

a. $2 \div \dfrac{1}{3} = \underline{6}$ Check: $\underline{6} \times \dfrac{1}{3} = \dfrac{6}{3} = 2$	b. $1 \div \dfrac{1}{4} = \underline{4}$ Check: $\underline{4} \times \dfrac{1}{4} = \dfrac{4}{4} = 1$
c. $6 \div \dfrac{1}{3} = \underline{18}$ Check: $\underline{18} \times \dfrac{1}{3} = \dfrac{18}{3} = 6$	d. $5 \div \dfrac{1}{4} = \underline{20}$ Check: $\underline{20} \times \dfrac{1}{4} = \dfrac{20}{4} = 5$
e. $5 \div \dfrac{1}{3} = \underline{15}$ Check: $\underline{15} \times \dfrac{1}{3} = \dfrac{15}{3} = 5$	f. $6 \div \dfrac{1}{2} = \underline{12}$ Check: $\underline{12} \times \dfrac{1}{2} = \dfrac{12}{2} = 6$

2. a. 18   b. 36   c. 32
   d. 50   e. 28   f. 40

3. a. 6 m ÷ (1/2 m) = 12. You can get 12 pieces.
   b. 2 c ÷ (1/4 c) = 8 servings of almonds.
   c. 5 kg ÷ (1/10 kg) = 50; He would need 50 tomatoes.
   d. 4 in ÷ (1/8 in) = 32 erasers

4. a. 5   b. 17   c. 5
   d. 2   e. 1/2   f. 6

5. a. 58/48 = 1 5/24   b. 22/48 = 11/24
   c. 4 9/14   d. 7 3/35

6. In one day, she walks 3/8 + 3/8 = 6/8 = 3/4 mile. In five days, she walks 5 × 3/4 mi = 15/4 mi = 3 3/4 miles.

7. You need (1 1/2) × (2 1/4 C) = 3/2 × (9/4 C) = 27/8 C = 3 3/8 cups of flour.

8. The answer to the multiplication (3/5) × (1/2) tells you what part of the original land gets used for growing crops (3/10).

Dividing Fractions: Summary, pp. 47-48

1. a. Each child gets ½ an apple. The division is: $10 \div 20 = \dfrac{10}{20} = \dfrac{1}{2}$.

   b. Each child gets 1¼ apples. The division is: $25 \div 20 = \dfrac{25}{20} = \dfrac{5}{4} = 1\dfrac{1}{4}$.

   c. He could see 5 patients in 3 ¾ hours. The division is: $1\dfrac{3}{4} \div \dfrac{3}{4} = \dfrac{15}{4} \times \dfrac{4}{3} = 5$.

   d. A half cup of almond flour goes into each batch. The division is: $1\dfrac{1}{2} \div 3 = \dfrac{3}{2} \times \dfrac{1}{3} = \dfrac{1}{2}$.

   e. Each person gets 4/5 of a pie. The division is: $2\dfrac{2}{5} \div 3 = \dfrac{12}{5} \div 3 = \dfrac{12}{5} \times \dfrac{1}{3} = \dfrac{4}{5}$.

   f. Six batches can be made with 3 cups of butter. The division is: $3 \div \dfrac{1}{2} = 3 \times 2 = 6$.

   Five batches can be made with 2 ½ cups of butter—just one batch less than with three cups (each batch takes ½ cup).

2. a. 50   b. 4/5   c. 2/9
   d. 1/20   e. 3 3/4   f. 48
   g. 2/5   h. 50   i. 1/60

Dividing Fractions: Summary, cont.

3. There are 10 such stretches: 2 1/2 mi ÷ (1/4 mi) = 5/2 × 4 = 20/2 = 10.

4. Answers will vary. Check the student's answers. For example:

   a. How many 1/2-cup servings can you get from 2 cups of ice cream? 2 C ÷ (1/2 C) = 4. You can get 4 servings.

   b. There was 1/3 of a blueberry pie left. Sylvia and Jessie shared it evenly.
      Each got 1/6 of the original pie. (1/3) ÷ 2 = 1/6.

5. a. 24 in ÷ (1/8 in) = 24 × 8 = 192 beads.
   b. Half as many, or 96.
   c. She will need 64 "S" beads and 64 "SS" beads. The pattern of SS-S-SS-S is 1/8 in + 1/4 in + 1/8 in + 1/4 in =
      3/4 in long. Two repetitions of the pattern is 1 1/2 inches long and four repetitions is 3 inches long. Now, the
      total length of the necklace was to be 24 inches, and 3 goes into 24 evenly, 8 times. This means we take
      8 × four repetitions, or 32 repetitions, of the pattern SS-S-SS-S. Since each type of bead is used twice in the
      pattern, that requires 64 of each kind of bead.

Dividing Fractions: The Shortcut, pp. 49-51

1.

a. $30 \div 5$	b. $\frac{1}{9} \div 3$	c. $\frac{1}{4} \div 2$
↓ ↓ ↓	↓ ↓ ↓	↓ ↓ ↓
$30 \times \frac{1}{5} = 6$	$\frac{1}{9} \times \frac{1}{3} = \frac{1}{27}$	$\frac{1}{4} \times \frac{1}{2} = \frac{1}{8}$
d. $\frac{1}{7} \div 3$	e. $32 \div 8$	f. $\frac{1}{5} \div 4$
↓ ↓ ↓	↓ ↓ ↓	↓ ↓ ↓
$\frac{1}{7} \times \frac{1}{3} = \frac{1}{21}$	$32 \times \frac{1}{8} = 4$	$\frac{1}{5} \times \frac{1}{4} = \frac{1}{20}$

2. Note that 1 is its own reciprocal, because 1 × 1 = 1. Or, you can write 1 as a fraction like this: 1/1. When you flip it,
   you get the same: 1/1 = 1.
   a. 6   b. 100   c. 1/5   d. 1/21   e. 8/7 or 1 1/7
   f. 3/11   g. 42   h. 1   i. 1/13   j. 6/5 or 1 1/5

3. a. 1/12   b. 1/40   c. 1/90
   d. 2/5   e. 1/8   f. 2/15

4. a. 16   b. 21   c. 55
   d. 1 3/5   e. 1/27   f. 9/10
   g. 7 1/5   h. 3 1/3   i. 8/35

Puzzle corner.
a. (29/10) × (5/7) = 29/14 = 2 1/14
b. (15/8) × (4/3) = 5/2 = 2 1/2
c. (23/6) × (2/3) = 23/9 = 2 5/9

Review, pp. 52-55

1.

$=$

      a. $\dfrac{9}{12} = \dfrac{3}{4}$

b. $\dfrac{56}{49} = \dfrac{8}{7}$

d. $2\dfrac{72}{84} = 2\dfrac{12}{14} = 2\dfrac{6}{7}$

c. $3\dfrac{15}{35} = 3\dfrac{3}{7}$

e. $\dfrac{12}{100} = \dfrac{3}{25}$

2.

a. $3 \times 1\dfrac{1}{3} = 3\dfrac{3}{3} = 4$

b. $2 \times \dfrac{5}{6} = \dfrac{10}{6} = 1\dfrac{4}{6} = 1\dfrac{2}{3}$

3.

a. $\dfrac{\overset{1}{\cancel{7}}}{\underset{2}{\cancel{14}}} \times \dfrac{\overset{1}{\cancel{3}}}{\underset{4}{\cancel{12}}} = \dfrac{1}{8}$

b. $\dfrac{\overset{1}{\cancel{5}}}{\underset{2}{\cancel{24}}} \times \dfrac{\overset{1}{\cancel{12}}}{\underset{6}{\cancel{30}}} = \dfrac{1}{12}$

4. a. 2 4/5   b. 5/21
    c. 17 1/5   d. 6 11/18

5.

a. Side lengths: $\dfrac{2}{3}$ m and $\dfrac{3}{4}$ m

    Area: $\dfrac{2}{3}$ m $\times \dfrac{3}{4}$ m $= \dfrac{6}{12}$ m$^2 = \dfrac{1}{2}$ m^2

b. Side lengths: $\dfrac{4}{5}$ km and $\dfrac{3}{4}$ km

    Area: $\dfrac{4}{5}$ km $\times \dfrac{3}{4}$ km $= \dfrac{12}{20}$ km$^2 = \dfrac{3}{5}$ km^2

6.

a. $\dfrac{5}{6}$ m $\times \dfrac{1}{2}$ m $= \dfrac{5}{12}$ m^2

b. $\dfrac{2}{3}$ in $\times \dfrac{1}{6}$ in $= \dfrac{2}{18}$ in$^2 = \dfrac{1}{9}$ in^2

7. $5 \times (3/4$ mi$) = 15/4$ mi $= 3 \ 3/4$ mi

8. $(2/3) \times (12/20) = 24/60 = 2/5$. This is what fraction of the *original* cake Sam ate. You can also reason it this way: There are 12 slices out of 20 left. Sam eats 2/3 of them, or 8 slices. And, 8 slices out of 20 = 8/20 = 2/5.

9. a. The dog ate **8 pieces** (2/3 of the 12 pieces).
    b. Four pieces are left.
    c. 4/20 = 1/5 of the pie is left now.

10.

a. $1 \div 3 = \dfrac{1}{3}$	b. $\dfrac{1}{2} \div 3 = \dfrac{1}{6}$

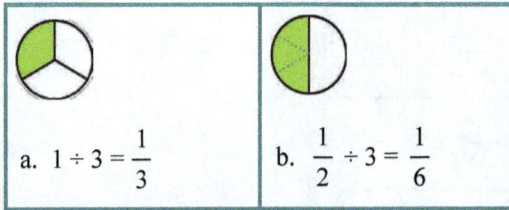

11. a. 6   b. 16   c. 1/10
    d. 1/21   e. 2 1/4   f. 1/16
    g. 2/3   h. 1/5   i. 1/4

12. a. Each piece is 1 3/4 inches long: 7 in ÷ 4 = 7/4 in = 1 3/4 in.
    b. (4/5) × (1/3) = 4/15
    c. 11 lb ÷ 5 = 11/5 lb = 2 1/5 lb
    d. (3/4) × $24 = $18

13. a. <   b. >   c. =

14. 1 − 1/3 − 1/4 = 1 − 4/12 − 3/12 = 5/12. So, 5/12 of the cake was decorated with strawberry frosting.

15. After a day, 5/6 of 30 slices were left, which is 25 slices.
    The family ate 1/5 of 25 slices, which is 5 slices.
    Therefore, 20 slices are now left.

16. Note: 9/16 cup or 9/16 teaspoon is not commonly found on
    measuring spoons. You would just use a tad over ½ cup or
    ½ teaspoon.

Brownies

2 1/4 cups sweetened carob chips
6 tablespoons olive oil
2 small eggs
3/8 cup honey
3/4 teaspoon vanilla
9/16 cup whole wheat flour
9/16 teaspoon baking powder
3/4 cup walnuts or other nuts

17. a. The 6 ½ in by 8½ in sheet has a greater area.
    The area of the 6 ½ in by 8 ½ in sheet is (6 1/2 in) × (8 1/2 in) = (13/2 in) × (17/2 in) = 221/4 in^2 = 55 ¼ in^2.
    The area of the 5 ¾ in by 9 in sheet is (5 3/4 in) × 9 in = (23/4 in) × 9 = 207/4 in^2 = 51 ¾ in^2.

    b. It is 3 ½ square inches larger in area. Subtract 55 ¼ in^2 − 51 ¾ in^2 = 3 ½ in^2.

Puzzle corner. a. b. d. e. are equivalent to 5/8.

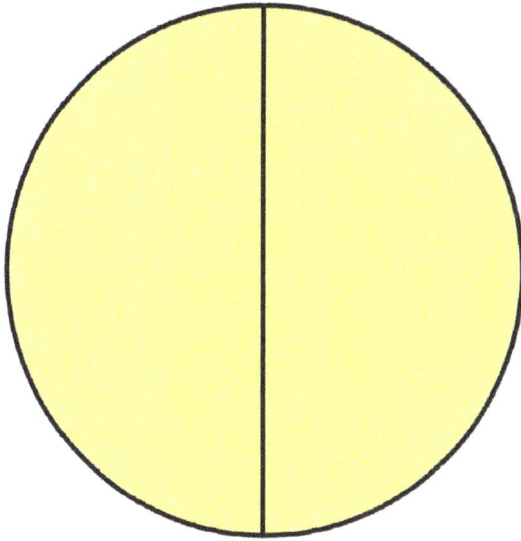
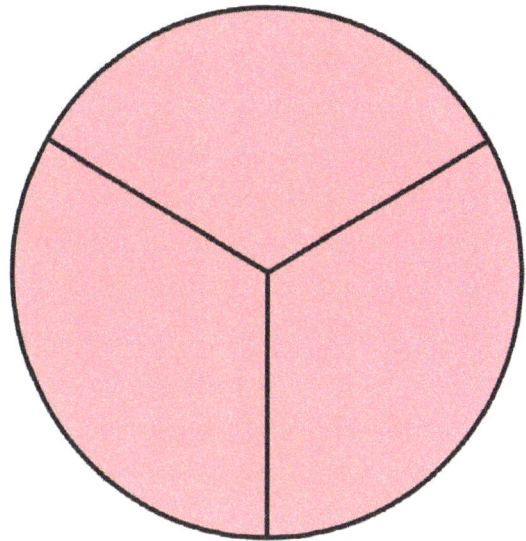
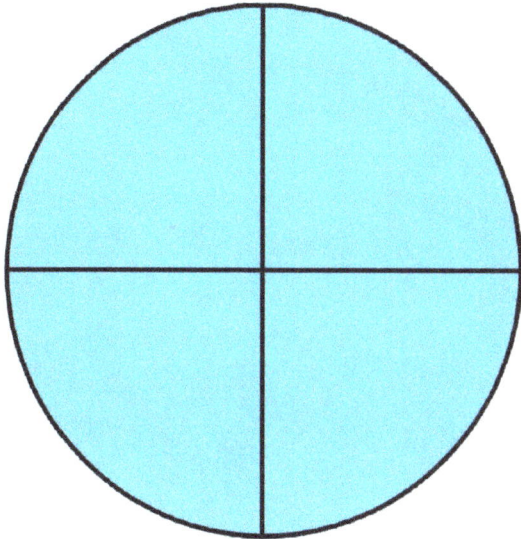
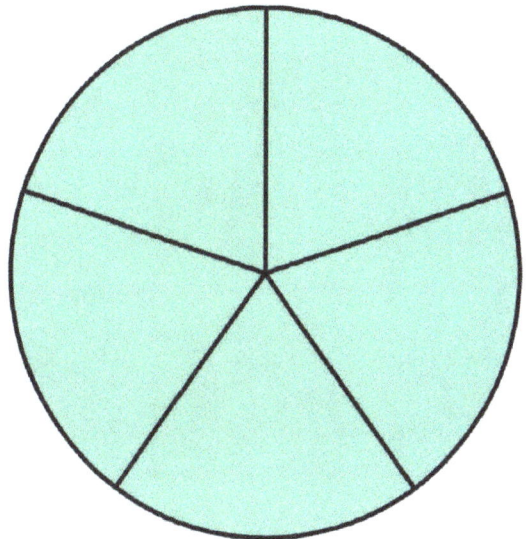
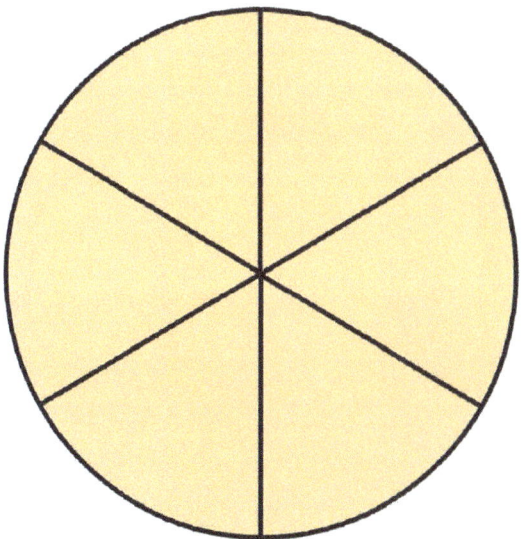
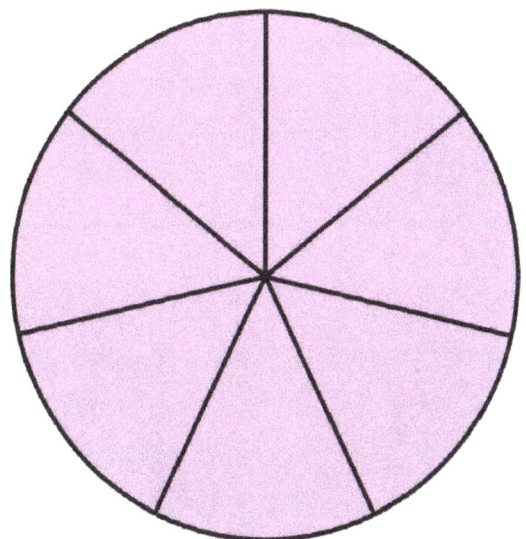

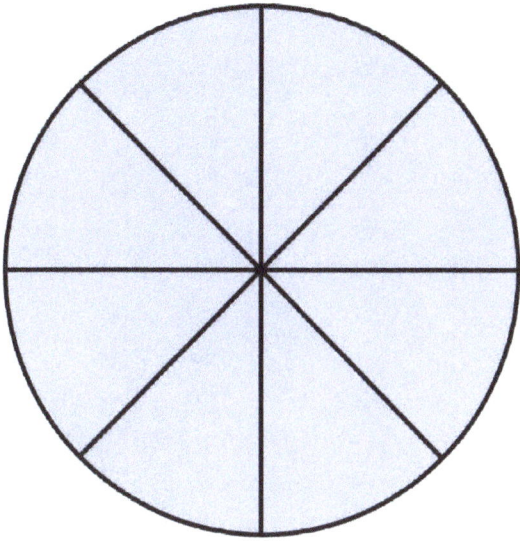

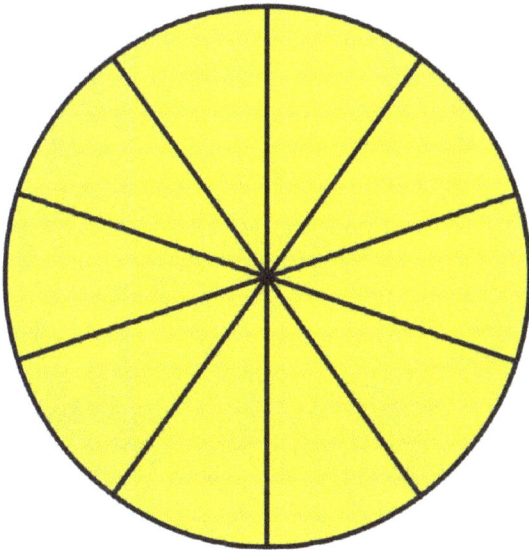
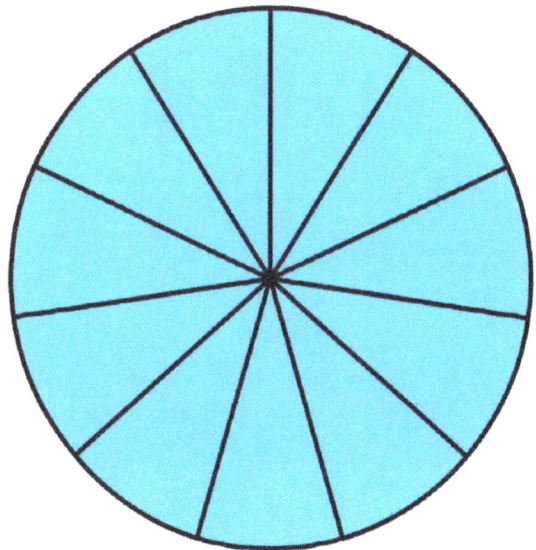
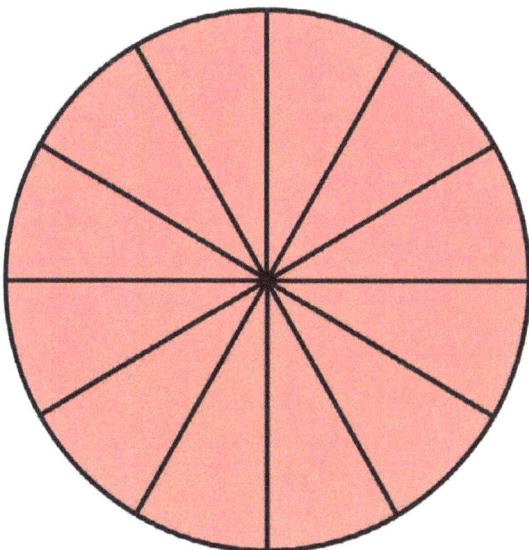
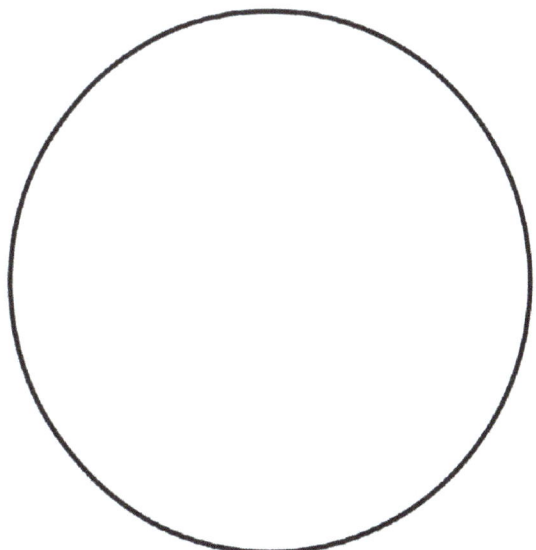

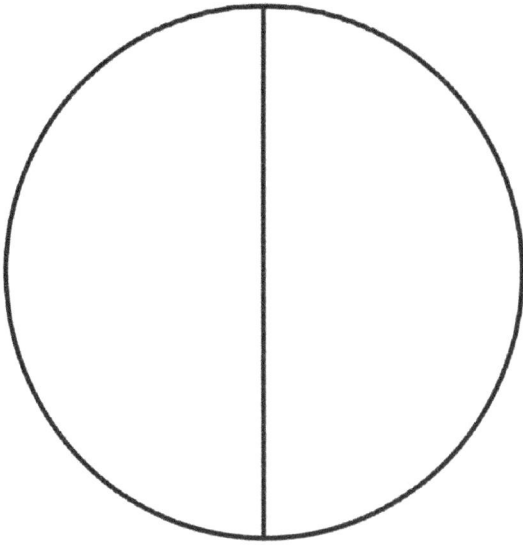
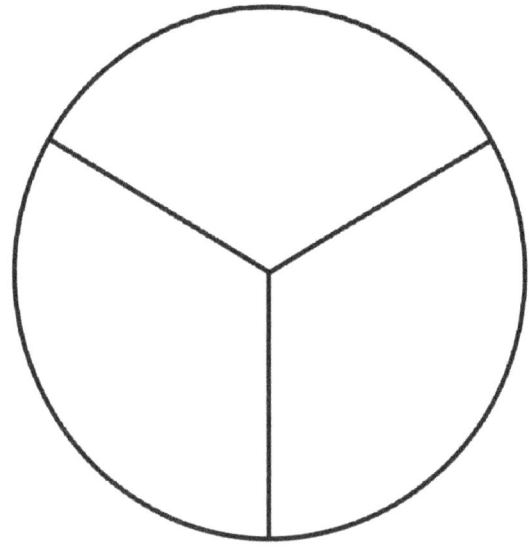
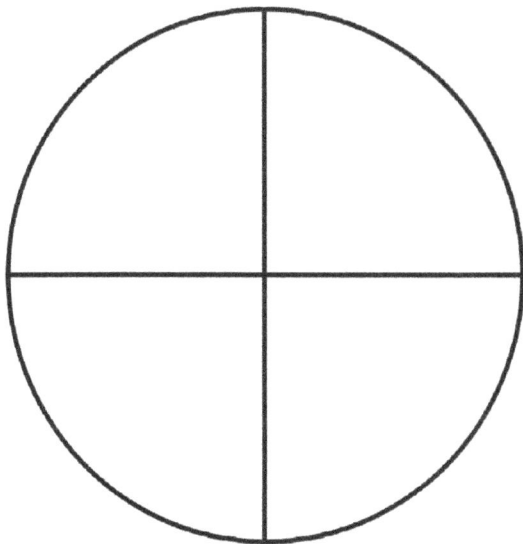
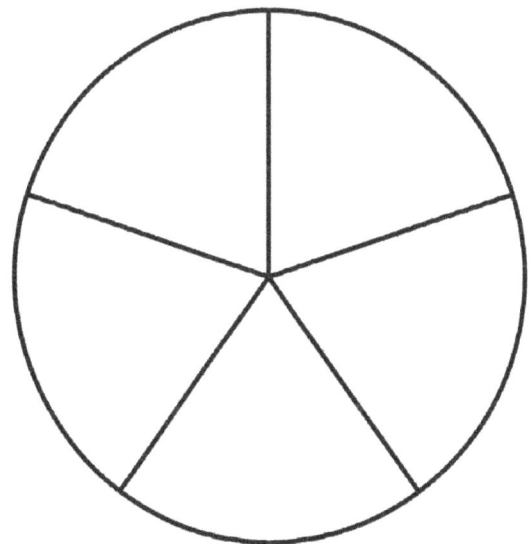
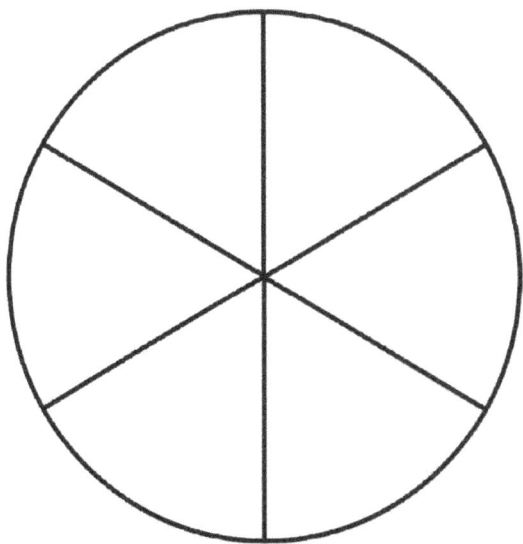
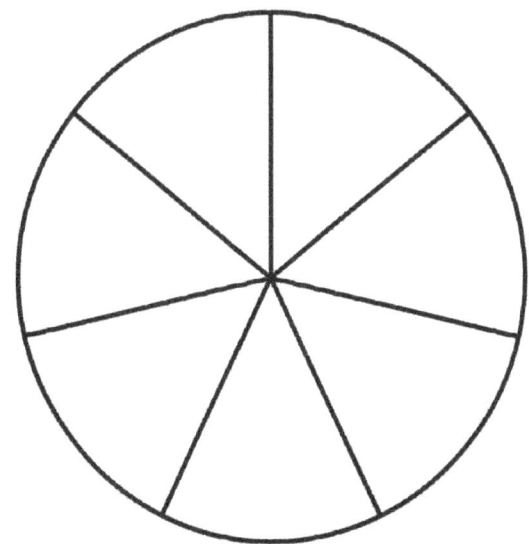

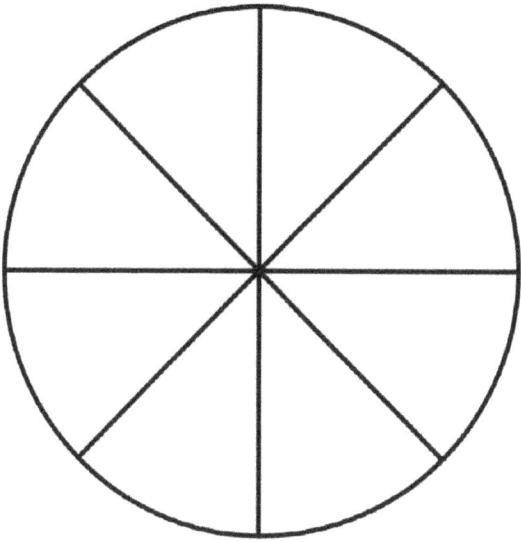
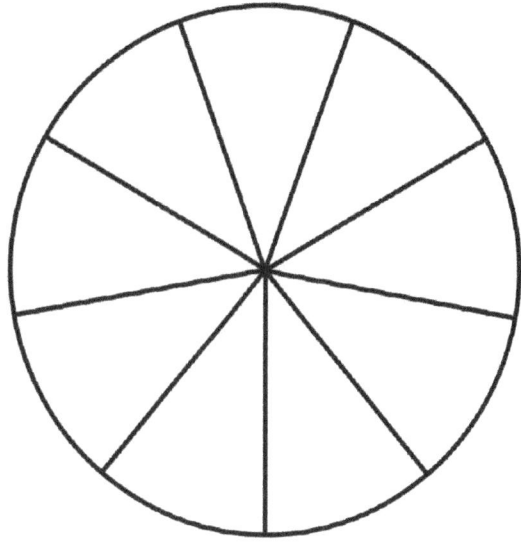
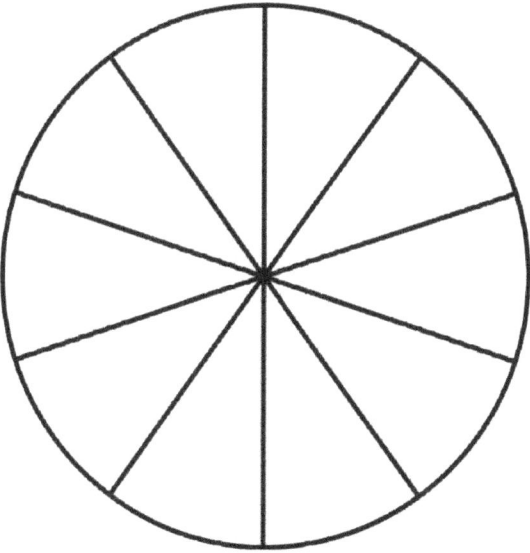
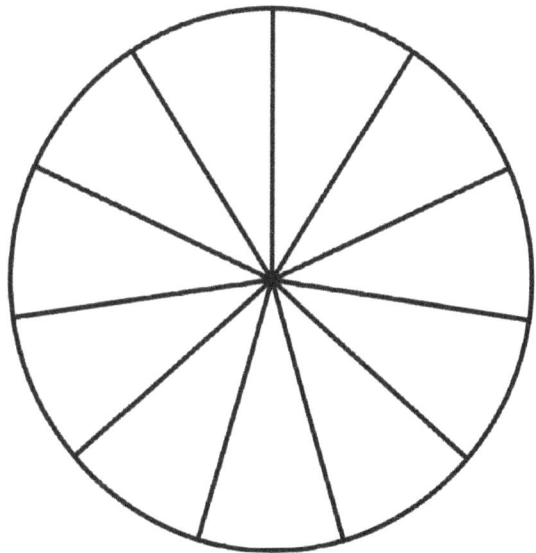
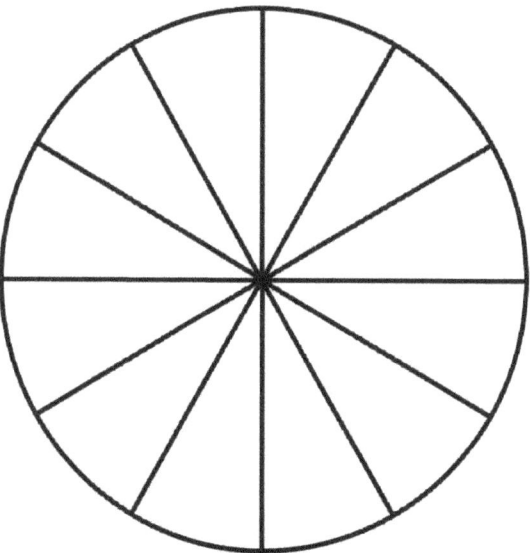
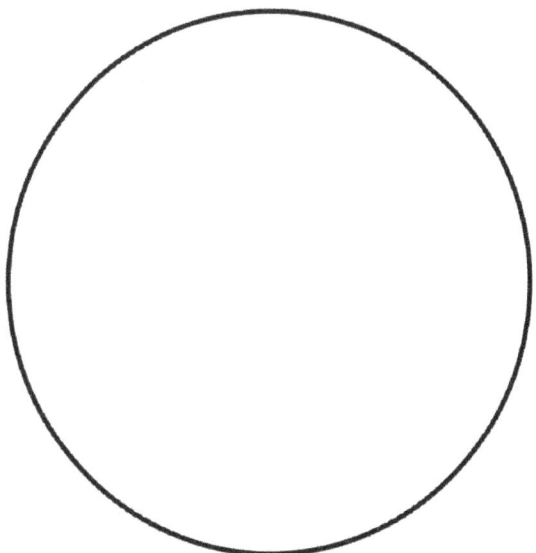

More from math MAMMOTH

Math Mammoth has a variety of resources to fit your needs. All are available as economical downloads, and most also as printed copies.

- **Math Mammoth Light Blue Series**
  A complete curriculum for grades 1-7. Each grade level includes two student worktexts (A and B), which contain all the instruction and exercises all in the same book, answer keys, tests, cumulative reviews, and a worksheet maker. International (all metric), Canadian, and South African versions are also available.

  https://www.MathMammoth.com/complete-curriculum

  https://www.MathMammoth.com/international/international

  https://www.MathMammoth.com/canada/

  https://www.MathMammoth.com/south_africa/

- **Math Mammoth Skills Review Workbooks**
  These workbooks are intended to be used alongside the Light Blue series full curriculum, and they provide additional review to the topics studied in the main curriculum, in a spiral manner.
  https://www.MathMammoth.com/skills_review_workbooks/

- **Math Mammoth Blue Series**
  Blue Series books are topical worktexts for grades 1-7, containing both instruction and exercises. The topics cover all elementary mathematics from 1st through 7th grade. These books are not tied to grade levels, and are thus great for filling in gaps.
  https://www.MathMammoth.com/blue-series

- **Make It Real Learning**
  These activity workbooks concentrate on answering the question, "Where is math used in real life?" The series includes various workbooks for grades 3-12.
  https://www.MathMammoth.com/worksheets/mirl/

- **Review Workbooks**
  Workbooks for grades 1-7 that provide a comprehensive review of one grade level of math—for example, for review during school break or summer vacation.
  https://www.MathMammoth.com/review_workbooks/

Free gift!

- Receive over 350 free sample pages and worksheets from my books, plus other freebies:
  https://www.MathMammoth.com/worksheets/free

Lastly...

- Inspire4 is an inspirational website for the whole family I've been privileged to help with:
  https://www.inspire4.com